ABERFAN

ABERFAN

A Story of Survival, Love and Community
in One of Britain's Worst Disasters

GAYNOR MADGWICK

WITH GREG LEWIS

First impression: 2016
Second impression: 2016
© Copyright Gaynor Madgwick, Greg Lewis and Y Lolfa Cyf., 2016

The publishers wish to acknowledge the support of
Cyngor Llyfrau Cymru

Cover photograph: Media Wales
Cover design: Y Lolfa

ISBN: 978 1 78461 275 7

Published and printed in Wales
on paper from well-maintained forests by
Y Lolfa Cyf., Talybont, Ceredigion SY24 5HE
website www.ylolfa.com
e-mail ylolfa@ylolfa.com
tel 01970 832 304
fax 832 782

Foreword by The Earl of Snowdon

VISITING ABERFAN IN the hours after the disaster was one of the most moving experiences of my life.

Gaynor Madgwick's book, *Aberfan*, is a brave, heartbreaking and inspiring journey in which she re-visits the story of what happened to her and to the whole community of Aberfan on that dreadful day.

It is a book that should be read by all of us in memory of those who died and those who survived.

Contents

Introduction by Vincent Kane

AT THE INQUEST into the deaths of 30 of the children killed at Aberfan in October 1966, as each child's name was read out, there were shouts of 'murderers'. As one child's cause of death was given as asphyxia and multiple injuries, her father called out, 'No sir, buried alive by the National Coal Board.' When the coroner remonstrated with him sympathetically, he persisted: 'I want it recorded. Buried alive by the National Coal Board. That is what I want to see on the record. That is the feeling of those present. Those are the words we want to go on the certificates.'

A few months later, after the tribunal of inquiry had found the Coal Board totally responsible for the Aberfan disaster and the deaths and destruction it caused, and had most severely criticised the Chairman of the Board, Lord Robens, the disgraced chairman offered his resignation to the Minister for Fuel and Power and, through him, to the Prime Minister. However, the South Wales Miners, along with their National Union, petitioned the government to reject the resignation and keep Lord Robens in post.

How could that be? It still seems extraordinary at half-a-century's distance but I believe that the balance – or rather imbalance – between these two conflicting points of view is the abiding conundrum of Aberfan – the riddle at the heart of the disaster itself and the series of shameful betrayals which followed it. The tip slide robbed the village of half of its children; the manoeuvrings over the following years of the various organizations which might have – and ought to have – brought succour to the bereft community robbed them of natural justice.

The tribunal report said there were no villains at Aberfan. Yes there was, there was one big villain. Coal. King Coal to which we all paid grateful homage in Wales for most of the 20th century. It was coal and the determination to keep producing it at all costs which caused the tip slide; it was coal which killed the children and it was coal and the desperate fear of losing it which prompted the dereliction of duty before the disaster and the cover-ups and half-truths which followed. When Robens was appointed chairman by Macmillan in 1960, he told the Prime Minister the state of NCB's finances made it next to impossible ever to make a genuine profit, but Supermac was unflappable. 'Don't worry, dear boy,' he said. 'Just blur the edges – just blur the edges.' And that is what Robens set about doing. He was an able man, had an iron will, a dominant personality and a natural leader; he quickly became known throughout the industry as Old King Coal.

One big problem for coal in the 1960s was oil, which was plentiful and very cheap; you could fill the tank of your Austin Mini or Hillman Imp or Ford Escort and have change for a pound note. More worryingly, industry was wallowing in the stuff, too. Another big problem was that there were – in the view of both the Macmillan and Wilson governments – too many pits and too many of them, especially in south Wales were ageing and creaking having been cut in the nineteenth century. But the third problem was the most politically sensitive of all – the miners and their union, the dreaded NUM, the most militant of them all. Start closing pits and they'd bite your hand off.

So Robens set about blurring the edges. He formed a close relationship – they called it a partnership – with Will Paynter, communist miners' leader of the South Wales Miners, then newly-elected leader of the National Union of Mineworkers, in which they agreed that the only long-term hope of salvation for the coal industry was a drastic reduction in the number of pits and, as a consequence, the number of miners. That was the policy and Paynter, a senior officer in the International

Brigades in the Spanish Civil War, was totally committed to it. He carried the union executive with him including Will Whitehead, another communist, who had succeeded him as leader of the South Wales Miners. The pits were to be closed one by one – doing good by stealth – rather than in one fell swoop which would surely have triggered a miners' revolt. As it was there was a spate of unofficial strikes; the 1960s were plagued with them. As a young television reporter I cut my teeth reporting from various collieries in the coalfield brought to a standstill at a moment's notice. I recall one occasion when the producer told me with a funny look as I set off, that London wanted to use my piece in the six o'clock news but had asked could I 'get some shots of miners singing as they emerged from the pit cages'. Tom Jones was already up and running; perhaps a few years later they might have obliged with 'Delilah'.

When Robens (and Paynter) took office there were 698 pits and 583,000 miners. When Robens left ten years later there were 292 pits and 283,000 miners. Job done? The government thought so, especially since productivity had increased by 70 per cent, which had enabled them to stop subsidies to ailing pits. Robens was a hands-on chairman. He insisted on getting out and about; in particular he visited collieries, he set a target of one colliery visit a week. He visited 350 pits in ten years; that's one every ten days. But he closed one every nine days; 400 in all. And nobody noticed. Except the miners in the pits which were closed. And crucially the miners who were fearful their pit might be next. Which leads us to Merthyr Vale colliery.

Merthyr Vale, with its seven tips, six of them pensioned off but No. 7 still tipping full tilt, circled the village of Aberfan like seven pillars of sombre unwisdom. Four of them, including No. 7, were perched on the sloping hillside; a policy described as unwise in what precious little national guidance there was on pit spoilage and tipping policy. They were all also built on water, either on something called the Brithdir water line or, in the case of No. 7, on two underground springs which were clearly shown on Ordnance maps. For 30 years or more the

11

streets and homes of the village were flooded, often knee-deep, and angry letters flew between the Borough Council and the NCB. Flooding is one thing. Tip slides are another. There had been two of them at Aberfan prior to 1966, one in 1944 and one in 1963, both of them, like the disaster in 1966, resulting from 'the fundamental mistake of tipping over surface streams and springs or seepages from permeable strata forming the sloping hillsides without taking any preliminary drainage measures' – so the tribunal was told. A soil mechanics expert told the hearing that with tipping, water is the source of all evil. It must not be allowed to get into the base of a tip. Failure to prevent that by proper drainage measures was the real explanation for the disaster.

The 1963 slide at the dormant No. 4 tip was a serious one; in fact it was a dress rehearsal – almost an exact copy of the disastrous slide of No. 7 three years later. But the NCB, that is to say the ten area, group, and divisional engineers, colliery and production managers named, blamed, and shamed by the tribunal refused to take it seriously; indeed some of them refused to recognise that it had happened at all for two years. Merthyr's Borough engineer sent out a round robin letter at one point headed, 'Danger from coal slurry being tipped at the rear of Pant Glas School', but it evoked little or no response. The appalling inaction, irresponsibility and failure to communicate still take one's breath away 50 years on.

Consider Her Majesty's Inspectorate of Mines and Quarries, which had a divisional office in Cardiff. No inspector visited the Merthyr Vale colliery tip complex at Aberfan for any purpose in the four years before the disaster, a period which included the big slide in 1963. This next fact is unbelievable but true. The senior inspector at the Cardiff divisional office of the Inspectorate of Mines and Quarries, who appeared before the tribunal, had been an inspector at the same Cardiff office 22 years earlier when the first big tip slide occurred at Aberfan in 1944. He confessed to the tribunal that the first time he heard of it was in 1966. Twenty miles up the road!

This inspector of mines and quarries must have driven past it umpteen times but he had not the faintest idea what had happened there.

Sir (later Lord) Edmund Davies, who chaired the tribunal with consummate skill, grasped the heart of the matter. I quote just three sentences which seem to contain all that posterity needs to know about Aberfan.

The stark truth of the tragedy flowed from the fact that notwithstanding the lessons of the recent past, not for one fleeting moment did many otherwise conscientious and able men turn their minds to the problem of tip stability. The incidents preceding the disaster should have brought home vividly to any having interest in coal that tips placed on hillsides can and do slip and having started can move quickly and far, so it was necessary to formulate a system aimed at preventing such a happening – to issue instructions, disseminate information, train personnel, inspect frequently. There was ample time for all this to be reflected upon and realised and effective action taken, but the bitter truth is they were allowed to pass unheeded into the limbo of forgotten things.

No chance of Lord Davies following unheeded into that limbo after such an epic judgment, of Denning-like proportions! And yet, and yet, the question which arises at this distance of time is why. Why did these conscientious and able men act or fail to act individually and collectively in this calamitous fashion? The reasons he gave for the ten individuals he named and blamed were bungling ineptitude in tasks for which they were totally unfitted, failure to heed clear warnings, total lack of direction from the top. He said they were not villains but decent men led astray by foolishness or ignorance or both. That in all conscience, he said, is a burden heavy enough for them to have to bear, without the additional brand of villainy. The iron fist in the velvet glove? It looks that way. Notice that he has added another epithet to his description of them. Not just conscientious and able, but decent, conscientious and able.

Then how the hell did it happen? How did 144 people,

including 116 children, come to lose their lives? Because, I believe, the learned judge omitted one reason, one vice from the list he tabled and it was the most glaring vice of all. Cowardice. Moral cowardice. They failed to look, they failed to report, they failed to question – these decent, conscientious and able men – because they were afraid or half-afraid of what they would see, of what they would hear, of what they might be required to do. They dared not even talk to each other about it because they knew intuitively that there was something wrong with tip No. 7. They were aware of the fears expressed time and again by the villagers, by the Borough Council. In January 1964 the *Merthyr Express* reported a meeting of the Town Planning Committee quoting verbatim from the minutes. Councillor Mrs Williams said: 'There are dangers from surface tipping. We had a lot of trouble from slurry causing flooding, but if the tip moved it could threaten the whole school.' Some of the ten, even just one of them, must have read or been made aware of that newspaper report. So serious, so startling, so threatening would it have been for the NCB to be criticised in public in this way that somebody – anybody – would have been delegated to check it out in order to deny it.

But no. At the back of their minds lurked the fear that if they looked, or asked, the answer they might be given or the evidence they might see would compel them to set in motion a process that would inevitably lead to the cessation of tipping, which would lead to an immediate cessation of production, which could well lead to a rapid closure of the colliery. To act or not to act; that was the question. To act was to put the existence of the pit in peril, so it was better not to act – that was the answer. The pit depended on the tip. No tip, no pit. The colliery manager wrote to the council in an argument about the introduction of tipping 'Tailings', and in his letter he warned that any threat to tipping at Merthyr Vale was a threat to the future of Merthyr Vale. In 1965 the NCB applied to the council for planning permission to divert some overhead lines at the tipping facility at Merthyr Vale colliery. The senior coal board official who wrote the accompanying letter concluded with this: '... If consent is not granted the tipping life of the area will be curtailed with a possible similar reaction on the life of the colliery.'

The closure policy of Robens/Paynter was 'full speed ahead' by the mid-1960s. It was ruthless; once a colliery was identified as being unproductive or uneconomic it was closed. No argument. In the Rhondda pits were dying like flies, so fast that the BBC commissioned me and producer Gethin Stoodley Thomas to make a television series that would capture something of the great coal tradition of the Rhondda, of the mines and the miners who made Rhondda a word that rang around the world, and to make the series before it vanished completely, and we were only just in time. It was called *The Long Street* and these days it is regarded as something of a history book. The four programmes went out in the spring and early summer of 1966 and were well received. We liked to think at the time they generated a feeling of national pride in what had been the warm relationship between coal and, not just the Rhondda, but Wales and Welshness. Three months later, Aberfan killed any such sentiment stone dead.

Uncertainty and insecurity was rife throughout the Welsh coalfield. Will we be next? Where and when will the axe fall? In 1963 Will Whitehead gave an assurance to the anxious miners and officials at Merthyr Vale that the colliery was not on the list for closure, but that was the first they knew that there was such a list and they must have wondered, given the problems facing the pit, how long it would be before they were added to it. As Alun Talfan Davies QC told the tribunal, 'accepting that in 1963 there was no intention to close Merthyr Vale, two things need to be said. Without the tipping facilities available on Merthyr Mountain the future of the colliery was to some extent endangered or imperilled, and having regard to an accelerated process of closures in south Wales there might well be an overriding fear that disaster might descend upon the village.'

Coal Board witnesses appearing before the tribunal faced a catch-22 dilemma. If they gave even the slightest indication that they had been worried about the stability of the tip then they were condemned out of their own mouths: 'you knew and you did nothing'. But if they denied any such knowledge they would be, as they were, criticised for ineptitude, foolishness, ignorance, etc. Perhaps they had no choice because officially the policy of the NCB, initiated and stubbornly maintained by Lord Robens from the moment he first appeared at Aberfan (36 hours late) till he was summoned before the tribunal on its 74th day, was to deny responsibility for what was a phenomenon of nature – a combination of heavy rain and unknown underground springs beyond human control. An act of God? Well, an act of Robens, more likely and that was not quite the same thing. So the decent, conscientious, able managers and engineers all towed the party line on the stand.

One witness, however, swam against the tide. One witness and one alone had the courage or the temerity to assert that he had thought the tip could slide, and that the slide could threaten life, but that he had taken no action. This was the Member of Parliament for Merthyr, Mr. S.O. Davies, known to all and sundry, far and wide, simply as S.O. Now S.O. was as old as Methuselah, as stubborn as a mule, and as tough as old boots. He was a 'miners' MP' and had been for 30 years and claimed he knew the local coalfield better than anybody, which I think was probably true. Nobody knew how old he actually was. When I interviewed him in his house I repaired to the kitchen with Mrs S.O. for a cup of tea while he 'put his collar on' for the cameras. I asked Mrs S.O. how old S.O. was and she said 'well I'm not exactly sure' and then, as I was turning away, she added 'and I don't think S.O. is exactly sure either'. A few years after the disaster the local Labour party dropped him as their candidate in the 1970 election because, they said, he was 'too old'. So he stood as an Independent and was re-elected, much to the delight of those of us who could tell a sure-fire winner when we saw one.

His testimony is important; crucial, I have come to believe, to a true understanding of what I have called the conundrum of Aberfan. The tribunal report records that 'he thought tip 7 might not only slide but its sliding might reach the village, and that when he expressed this fear to miners in Aberfan they told him, "You make a row about that and what will happen? They will close the blessed colliery."' At this point Edmund Davies took over the questioning.

'You thought the slide might reach the village with a risk to life. Is that right?'

'Yes, certainly.'

'If you entertained substantial fear of risk to life, what did it matter if people asked you not to take steps? Why not take them – if there was a risk to life?'

'If I had taken them I have more than a shrewd suspicion the colliery would be closed.'

'So you went through a tortured process of thought, of weighing one against the other. The risk to life on the one hand, and the risk of colliery closure on the other. You came down on the side of taking no action which might risk colliery closure? Now think, before you answer Mr Davies. You understand it is a question of considerable gravity?'

'Yes I have thought. But I had to consider the general feeling of the mass of the people in that ward. But if I had had any official approach made to me about the tip – I should not like to tell the enquiry that we could have stopped it, quite frankly – but if I had been asked to do so, I would have done it.'

'Thank you.'

'Whatever the consequences for the colliery.'

S.O.'s evidence was strongly challenged by counsel for – not the Coal Board – but, significantly, the National Union of Mineworkers and it is worth hearing what he said. 'If his account is truthful – and I am not suggesting that he is deliberately untruthful – then he bears one of the largest personal burdens of responsibility for the disaster. He readily assumed, more than any other individual in the case, a knowledge of danger and absolute inactivity in dealing with it.'

He urged the tribunal not to accept S.O.'s evidence on the grounds that he didn't know what he was saying and remarkably that is what the tribunal did. 'He was the only witness to give such testimony and we doubt that he fully understood the grave implications of what he was saying. Were we convinced that he did – he could not escape censure.'

Take my word for it, in 1967 S.O. Davies was fully *compos mentis*. As he was when, three years later, entirely on his own, he took on the official Labour candidate in one of the safest Labour seats in the country, turned the vote around and beat him. The question that dances around the episode of S.O. Davies's testimony is not whether he understood the grave implications of what he was saying but whether they, the tribunal, understood the grave implications of what he was saying and if they did, is that why they closed their ears and refused to accept it? For the implication of what he was saying is not just that he knew the tip could slide and endanger life, but that they – the community of Aberfan – knew that the tip might slide and endanger life; they knew because he told them – not just the miners but 'the mass of people in the ward', and that is pretty grave. Should they not bear – what was it the NUM's counsel said, 'the largest burden of responsibility for the disaster, a knowledge of danger and absolute inactivity in dealing with it'? Not just inactivity, they positively urged him, pleaded with him not to do anything about it. It gets worse. He told the judge – who had just warned him it was a question of considerable gravity – that if he had had any official approach made to him about the tip he would have done it, i.e. taken steps. From what 'official' quarters might such an approach have come. Council officials? NUM officials? It seems highly unlikely that information as momentous as that he passed on to the 'mass of people' in the Aberfan ward would not have leaked out eventually to reach official ears, in which case why did those officials, whoever they were, not make the approach, which they undoubtedly should have, in order to save endangered lives? If foreknowledge coupled with inactivity was indeed the cardinal sin, as the tribunal believed, then that 'personal burden of responsibility' might have had to be shared round a lot more shoulders than those of the veteran MP. More cosy perhaps to write him off as being off his chump.

They knew. They all knew. The council. The Aberfan community, anybody who read the *Merthyr Express*, the school headmistress, the colliery employees who worked on top of the tip and who were still tipping until the tip started to go, and all the managers and engineers who were eventually pilloried by the tribunal. They all knew that there was a question mark over the safety of tip No. 7, but they preferred to look the other way, except those brave few like some of the councillors or the headmistress who did ask questions or raise complaints but who were fobbed off or ignored.

If it is true that it was fear; a fear that dare not speak its name which paralyzed the possibility of action, what was it that they were afraid of? Pit closure, which would have meant that a serious unemployment problem might blight the area. The fear was based on the widespread assumption that if tipping stopped, then production would stop and Merthyr Vale colliery would be added to that National Coal Board list of pits for closure which they had heard about in 1963. Was that a reasonable assumption? Yes I think it was, or it must have seemed so at the time. The National Coal Board, in partnership with the National Union of Mineworkers, was actively looking for pits to close in their bid to make the coal industry viable. In ten years they closed two-thirds of the collieries in Britain and in 1966 they were more than halfway through that cull. The south Wales coalfield had suffered more than others. Why should a pit where production had stopped and there was no quick way to start it again be spared?

The reasons, the specific reasons why that tip slid down the hillside at Aberfan in 1966 were exactly those which the tribunal so painstakingly and scrupulously enumerated. But if, with 50 years' hindsight we look for the cause – the underlying cause, and if we look with uncluttered minds now that coal is dead and gone and all the scars and monstrosities it inflicted on the landscape, not just of Aberfan but of Wales, have been cleaned and cleared, how can we fail to conclude that the underlying cause was the intense pressure brought to bear on a frightened coal mining community by the policy of widespread and rapid pit closures implemented by the National Coal Board, supported by the National Union of Mineworkers and two governments, Conservative followed by Labour, with the objective of making coal viable. That support by the trade union and by the Labour party in Wales and elsewhere persisted after the disaster. Read the Hansard report of the debate in the Commons on the tribunal report. One after another Welsh Labour MPs in mining constituencies spoke in mitigation of the 'wholly to blame' National Coal Board and urged that Robens who had submitted his resignation should remain in post. Leo Abse was the sole Labour exception. Emlyn Hooson (Liberal), Gwynfor Evans (Plaid Cymru) and David Gibson Watt (Conservative) wanted him to go but the Welsh Labour Party wanted him to stay on, and the trade union wanted him to stay on and everybody else who petitioned the Prime Minister wanted him to stay on because – never mind Aberfan, in spite of Aberfan – he was doing a good job, doing what had to be done and he was the only man who could do it. I haven't made that up, those are not my words, that was what they said. Documents published 30 years later showed that Robens orchestrated the campaign for Wilson to reject his resignation, didn't send it until he knew it would be rejected and even made the Minister of Fuel and Power, Richard Marsh, remove a sentence from the letter he would send back rejecting the resignation. Incidentally, the best speech made in the debate came from the Shadow Minister for Fuel and Power who just happened to be one Margaret Thatcher, who with great forensic skill ripped the

Coal Board and its chairman to pieces. She revealed one piece of chicanery which had escaped everybody's attention. At the time of the disaster the NCB's divisional chairman, Mr Kellet, was attending a power conference in Japan. He was given instructions to stay there and did not come back in time to give evidence to the tribunal. But part of the case advanced by counsel for the Board was that the absence of a tip policy should not be laid at the door of the Board but at the door of the division. Surely, said Mrs Thatcher, if someone is going to advance that argument he must ensure that the head of the division is in the country so that he can be brought before the tribunal to give evidence. Instead, he had ensured that Mr Kellet stayed out of the country.

The irony of all this is that both sides were proved to be wrong in the long run, the Coal Board and the Merthyr Vale community. Coal could not be made viable, no matter how many pits were closed. Dai Francis was, in the 1960s, secretary of the Welsh Miners' Union, grey and slow-spoken like a wise old owl. He too was a communist but he was also a Methodist; four parts Marxist, he told me, and one part Methodist. Like a dry martini, shaken but not stirred. Certainly not stirred because he had the great gift (for a trade unionist) of never losing his temper on air. I remember one TV debate about the merits of expensive coal versus cheap oil which he concluded with a smile and a shake of the head and, in that slow didactic manner with his strong, beguiling accent, he said, 'Vincent, the Arabs will not for ever live in tents.' Damn right they didn't. Not so many years later at about the time striking miners were winning the battle of Saltley Gates under the leadership of a young Arthur Scargill, OPEC came on stream and the price of oil went up 250 per cent in 18 months, throwing Western economies into a crisis which is still playing itself out. But even without cheap oil, coal was not competitive. Throughout the 1970s and up to and beyond the final and fatal miners' strike, coal from Poland and from faraway Australia, even with the costs of transport thrown in, was cheaper than British coal.

Robens had been right when he told the Prime Minister in 1960 that it was next to impossible for the Coal Board to make a profit. The illness, he and Paynter diagnosed, despite their drastic surgery, was terminal.

And the Merthyr Vale community were wrong too. When tipping stopped the pit was not closed; in fact it lasted longer than the National Coal Board which was wound up in 1987. Merthyr Vale was one of the last Welsh pits to go when it closed in 1989, 23 years after the disaster.

About a dozen years after the tribunal I made a profile of Lord Edmund Davies for *Week In Week Out*. He was retired, a widower, living 'over the shop' at Gray's Inn where we set up the cameras in his beautifully appointed rooms. Sadly, he had lost his vigour – he had been a tough cookie in his time; not long before the tribunal he was the judge at the trial of the Great Train Robbers where he handed round 30-year sentences to all the principal villains, but now he was old and rather frail, fully understanding but somehow gentle. Nevertheless, I reminded him of the men whom the report had named and blamed, 'not villains but decent men etc.' At the time I told him there had been much surprise and some anger that they had not been charged with manslaughter. He thought for a moment and then he said, 'But we thought they would be. We assumed they would be.' And that was that. He wouldn't be drawn any further. He may, of course, have been speaking with hindsight – there had been a considerable furore on the matter – but taking his words at face value what they mean is that this very senior and distinguished judge and his two expert lay colleagues, having examined the whole matter in detail for nearly three months, were of the view that there was enough prima facie evidence against certain of the witnesses to warrant a prosecution for manslaughter. Why was such a prosecution not initiated and who would have been responsible for the decision whether or not to initiate it? Who else but another Welsh MP for a mining constituency, the Attorney General Elwyn Jones, who had muzzled the press before the tribunal started and also given an assurance before it started that there would be no prosecutions and he officially ruled them out after the tribunal's report was to hand. We cannot say now that they were guilty of manslaughter nor could anybody at the time. Only a legal trial could establish that, but to deny such a trial was to deny the Aberfan community natural justice and that was a betrayal. The first betrayal of many as it turned out.

Lord Robens and the Coal Board betrayed the Aberfan community continuously. Robens fought tooth and nail to limit the cost of Aberfan to the NCB. The opening offer to each bereaved family was £50 and it wasn't increased to £500 and paid out until 1970 when Robens's 10-year stretch was over. He seized on a doubtful report that the remaining six tips were safe, as grounds for refusing to remove them. Instead he offered to 'contour' them and blend them into the slopes of the Merthyr Mountain; perhaps he had some kind of theme park in mind! Eventually forced to remove the tips, he persuaded the Prime Minister that the appeal fund should be raided for a contribution to the cost. The fund had reached £1.6m – about £28m today – and 'under intense pressure' the trustees were forced to surrender £150,000 (£3m today) which was 10 per cent of the fund. This was outrageous and probably illegal but the trustees, who were generally fairly useless anyway, just caved in. They should have threatened to resign en masse but they didn't and that was a betrayal of the Aberfan community. The Charity Commission, which many years later formally apologised, just nodded the outrage through and that was a betrayal of the Aberfan community. Indeed the Commission was hostile to the community from the start; at one stage they insisted that monies should be paid only to those parents of whom it could be established 'that they had been close to the child or children'.

The bereft community might have looked to the government, a Labour government, for support but they didn't find any when it mattered other than warm words of sympathy. In particular, they might have looked for support from the Secretary of State for Wales, George Thomas, who had taken over from Cledwyn Hughes. Wilson, the Prime Minister, was a pragmatist. Generally speaking he would do what he was able to do, what he could get away with. He would push at a door and if it gave he would go through it with his policy but if the door didn't give, if there was resistance he would stand back and think of something else. Wilson was aware that the decision to take the money was highly contentious; he would certainly have discussed it with his Welsh Secretary of State. That was what the Welsh Secretary was for – a funnel through which Welsh matters and problems were channelled into the cabinet and the Prime Minister, and, in reverse, government decisions and policies were channelled out to Wales. He spoke for Wales, for heaven's sake. Whether he demurred and argued with Wilson about it we do not know, but he accepted it and that was all the Prime Minister needed. George was all hearts and flowers in south Wales but he was true steel rock solid behind the Prime Minister in London. I thought then and I am convinced now that if he, the Secretary of State for Wales, had told Wilson that he would resign if this cruel and probably illegal policy went through, the Prime Minister would not have felt strong enough to do it. But he didn't, and from the Secretary of State for Wales that was a great betrayal of the Aberfan community. Thirty years later another, newly-appointed Labour Secretary of State for Wales, Ron Davies, redeemed the reputation of the office and of the party when he insisted that the money should be repaid to the fund and in 2007 the Welsh Assembly paid another £2m to bring it in line with inflation. For Ron Davies it was the honourable and – here's that word again – decent thing to do, and I salute him across the years.

But if I am scarifyingly honest, in one sense the media betrayed the Aberfan community and I was of the media and in quite a prominent role. Somehow or another, in that first five years after the disaster, as controversy followed controversy, though never directly articulated, a general climate of opinion developed in which the surviving community were seen to be 'the problem'. They were 'awkward' or 'greedy and grasping', there was talk of the appeal fund being the second Aberfan disaster, they were 'troublemakers', as I say they were 'the problem'. While the journalists, the press and broadcasters, didn't say as much, didn't light this fire of rumour, we fanned the flames insomuch as we didn't jump on it and smother it as robustly as we should. We didn't shout as loud as we could until Wales was deafened by our hammering home the simple truth that it was the trustees who were the problem, and the Charity Commission; it was the government and the politicians who were the problem, and most of all it was the National Coal Board who were getting away with murder and especially Alf Robens who were the problem. The Aberfan community were the victims, not the problem, and we were letting their tormentors, for such they were, off the hook. The press, the media, the fourth estate, has an abiding responsibility to probe and to penetrate. In the Aberfan period, perhaps Wales's darkest hour in the 20th century, we should have been passionate in pursuit of the truth. Instead we were pedestrian.

I have made no mention of the children. That is because I have found a special place for them which I urge you to share. I run a picture of them through my imagination, laughing and chattering as they leave the assembly and settle in their classrooms. Then I stop the picture at that moment; freeze it, do not proceed a second further, don't dare to go there. Freeze them in time like the young lovers on Keats's Grecian Urn – 'for ever panting, and for ever young'. Keats it was who said 'a thing of beauty is a joy for ever' and so they are; frozen in time at that moment, happy, innocent, and beautiful. Immortalize them like that and they will live for ever.

Like the children of Hamelin in Browning's 'Pied Piper' who left our world behind as they:

Tipping and skipping ran merrily after,
The wonderful music with shouting and laughter...
When, lo, as they reached the mountain-side,
A wondrous portal opened wide,
As if a cavern was suddenly hollowed;
And the Piper advanced and the children followed...

Did I say, all? No! One was lame,
And could not dance the whole of the way;
And in after years, if you would blame
His sadness, he was used to say, –
"It's dull in our town since my playmates left!
I can't forget that I'm bereft..."

Just such a one is Gaynor Madgwick who at eight years old was pulled injured from one of the classrooms where her friends died. She was left behind to live out her life. This is her story, sad, sweet, sentimental, and authentic. I commend it to you.

Preface

AT 9.15 ON the morning of Friday, 21 October 1966, the village of Aberfan heard a dreadful sound that was to haunt the community and a nation for a lifetime.

I was eight years old and had just sat down to a maths lesson at Pantglas Junior School.

Four years later I wrote a secret journal, trying to describe that moment. I wrote: 'I heard a terrible, terrible sound, a rumbling sound. It was so loud. I just didn't know what it was. It seemed like the school went numb, you could hear a pin drop. I was suddenly petrified and glued to the chair. It sounded like the end of the world had come.'

After days of bad weather, water from a spring had destabilized a huge coal slag tip – one of the black man-made mountains which surrounded the village – and thousands of tons of this heavy, dirty, deadly material were rushing down the hill towards us.

The sound got louder and louder. I remember thinking something terrible is going to happen. I got up from my chair and tried to run for the door. I never made it.

Then this hellish mass came through the window and everything literally went black.

When I woke I was inside a terrifying nightmare, pinned to the back corner of the classroom. Debris filled the room; there were children everywhere. I was crushed behind a large old radiator, which had been wrenched from his fittings. Its massive pipes had protected me from suffocation but had also smashed my leg.

A child's arm hung from the next classroom over my shoulder. I squeezed and pinched to see if it was alive, but there was no movement. For all I knew it was the hand of my brother.

Around me was a wall of sludge, five feet high. A couple of my classmates were lying underneath me and at the side of me. One little boy, a friend, died by my side.

*

For the past 50 years I have lived as a sort of prisoner or victim of my past. Now I am trying to break free.

I started this book by looking again at the writings of my young self. I've tried to explore the determination, courage and resilience which got me through.

Then, I set out on a journey, to find those same qualities in my community, to see how it had coped, survived and often thrived.

I've tried to tell this story in a way in which it has never been told before, beginning by reliving Aberfan through the eyes of a survivor. This traumatic experience of horror, helplessness and injury directly affected me, my friends and my relatives.

There was no time to prepare for such devastation and destruction, and afterwards there was a mixture of anger and injustice, which threatened to overwhelm our memories of the loved ones we had lost. It was not just a matter of terrible grief; Aberfan was a man-made disaster made worse by decisions which affected all those trying to grieve.

As a survivor, now 58 years old, I have been haunted by the memories of the Aberfan disaster. In my childhood and adulthood, I have had to cope with the severe psychological, physical and emotional consequences of that traumatic event. The overwhelming memories, thoughts and feelings and the lasting effects that continue to emerge from that day, still impact and disrupt my life and my family's lives.

But I needed to find out how others felt. I wanted to track down the people who saved and helped me that day, and talk to them.

I wanted to speak again to people forever linked by the disaster and those individuals whose kindness and compassion ensured I lived to have a fulfilled life. I needed to know how they have coped and if they too are haunted by their memories.

Time is running out. October 2016 marks the 50th anniversary of that awful day. For 50 years we have been trying to recover from the Aberfan disaster. It's a long road, and we take it one day at a time.

I wanted to create the fullest picture of the disaster and its aftermath while people were still around to tell their story.

On my journey I have met with doctors and reverends, royalty and politicians, miners and psychiatrists, always returning to my home, which is still in Aberfan, to take stock of what I have learnt.

For some, the past was truly buried, but I needed in some way to excavate their memories and recapture their feelings and emotions by asking questions that have been haunting me and perhaps them for half-a-century.

For me, I can't start the next chapter of my life if I keep re-reading the last one; this book will help me move on. My hope is that it will help others move on too.

This book is based on hours of conversation and research. All those I met gave their time freely and generously. I hope that their voices help make this book truly unique.

Chapter One

A Family from Aberfan

'My name is Gaynor Minett. I have blonde hair and hazel eyes. I live at 17 Bryntaf, Aberfan, which is a well-known village. I have five sisters, one brother and my parent...'

WHEN I WAS a small child I kept a diary. It began like that of any other young girl. But it was to be dominated by the Aberfan disaster. I was eight when I was dragged out of the ruins of my primary school. Eight when I lost my brother and sister. By the time I was in my early teens my private journal showed an obsession with seeking to explore how that one moment in time changed my life for ever, and those of the generations of villagers that would follow.

It was the day the world wished had never happened: the day hundreds of tons of colliery waste, which had been piled on a mountainside, slid downwards and engulfed the school in our mining village of Aberfan.

The scale of the tragedy shocked the world. One hundred and forty-four people died – 116 of them were my school mates.

My brother and sister were in classrooms either side of mine. We had all dressed for school together.

Only I would ever be coming home.

*

I was born Gaynor Minett on 12 March 1958. We lived in the Bryntaf area of Aberfan, opposite the quarry which my father,

Cliff Minett, owned. Dad had met my mam, Iris, when she was just 16. They had married two years later in Aberfan, with Mam one day shy of her 18th birthday. I had three sisters and one brother: Belinda, the eldest, Marylyn, Carl and Michele. I was the middle child. There was one more sister yet to come.

Mam worked hard to care for us all. She always kept us very clean and we had a nice home. We were a loving family, but having a large group of siblings all quite close together in age caused a few arguments and jealousy. As soon as one had something, we all wanted the same. Carl was very mischievous but always seemed to be spoilt by all his sisters. He swore and shouted at my mother and many a time had a good hiding.

My father had his own quarry business and got paid reasonably well. We were never short of money and had quite a reasonable house. However, it was too small for us: being a family of seven, we did not have enough room. We were always getting under each other's feet. My father dreamed of building a new bungalow for us all.

I went to Pantglas Junior School. For generations every parent in the village had sent their children there. It was a friendly place and I really loved it. Belinda went to school in Quakers Yard because she passed her 11+. Marylyn was a class higher than me and Carl was in the class lower. Michele was younger, and still at the Perth-y-gleision Infants' School.

I have strong memories of being five years old. I was a very quiet child and quite fragile. My mother dressed me in little tartan skirts and frilly blouses. My long, fine hair was worked into a style which my mother called 'a sausage' and she rolled it into this shape every morning. I was a pale girl and my main trouble was quickness to tears. I cried for the least little thing. My mother said I should have been called 'Bawler'. I remember when I was seven I had a terrible row with my mother and I stormed out and ran away for the day. My mother had to get everyone to look for me. I was hiding up on a farm in the hay and I stayed there for at least two hours while my parents were getting really worried. After a while I eventually came down and I found my parents. My mother hugged me and gave me another row, but she was sorry and I was sorry too for frightening her. I was the sort of child who did like to wander off. Sometimes I went down the park or for walks up the Canal Bank, which was so-called because the Glamorganshire Canal used to come through Aberfan.

I had many friends. My best friends were Sandra Leyshon and Desmond Carpenter. We went everywhere together and were like sisters and brother. Desmond lived three houses away from me. My sister, Marylyn, was two years older than me and she also liked Desmond. We fought over him, but she always won. My friends were all really great and I wouldn't part with them for the world. Sandra was like a sister to me. I went on holiday with her and had tea at her house and we were good companions. The streets every night would be filled with children and teenagers.

Gangs of us used to get together, at least 20 of us, and we'd pick teams and play jailer. This was a great game and we all enjoyed it. The girls ran off first and the boys would try to catch them. Many of us used to have a sly cuddle which we found made the game all the better. When we were caught, we had to be taken up the street to sit on the window sill and wait for someone to touch us out. This would be a hard time for the child who had to be the jailer. The trouble with this game was you had to run a lot and, as soon as you got exhausted, you'd get caught. But I was one of the fastest runners and took quite a long time to catch. Desmond would chase me around the street and two other boys would wait around the corner to jump out at me from an old set of garages. One night, I fell down a gully. I went head over heels. My head was cut and the graze on my leg was bleeding terribly. I had to go up to the hospital where they cleaned it and put a dressing on it. I still have the scar today.

I always had a problem falling down. My knees were scarred and scratched. You'd never see me without a bandage around my leg. A few months after my eighth birthday, I learned to ride a bike. I had a two-wheeler bike with small wheels and I thought I was great. It was on Sunday afternoon when the street held a meeting in the quarry to decide where we'd go for our summer day's outing. I had my bike and rode up and down the street. I was at the top of the street when a terrible thing happened to me. Somehow I hit the brakes hard. I went straight over the handlebars and knocked my head on the road. It knocked me out for a second and I got up completely dazed. I couldn't feel anything. Everyone came running and then it happened: blood came spouting out of my forehead. I was really petrified and I thought I'd die. The shop assistant, Mr Jones, a first-aid man, carried me over his shoulder, blood streaming from my head, and I fell unconscious. Mr Jones said I didn't need hospital attention. I was lucky. Everyone came in to see me and I was in a shocking state. I had to stop in for three weeks as my head thumped as I walked, but it finally got better.

During the time I was kept in the house, Sandra, my best friend, had an accident too. She fell through a window and was rushed to hospital with her head covered in blood. Everyone got anxious waiting for her to come back. They did not keep her in but she had a terrible headache and injury. On top of her eye was a lump the size of a golf ball and it was said to be a clot. She had to go up for daily treatment but she was all right. Her mother was worried about her. As soon as the two of us were well enough to go out, we went around together and were back to our old friendship again. But Sandra still had that clot on her head and doctors were getting worried because if it didn't burst she might die. Thankfully, Sandra and I didn't know this and our being together made her feel happy.

One night, we went to the fish shop and, as soon as we went down the steps, a dog came out and started barking at us. I ran and must have hit her accidentally on her lump and it burst. I can still see her screaming with fright and the blood pouring down her face. She ran home screaming and I was scared of what I had done. Her mother rushed her up to the hospital where they were more than pleased that it had burst. Her mother and father thanked me and, in a way, I felt I had saved her life. Sandra was praising me but I said that's what friends are for. It's funny what you think when you are child and have no understanding of what illness and death mean.

Just before Christmas 1965 we had the wonderful news that my mother was having a baby. As she was small, she didn't tell us until we could see. She was five months pregnant and had four months to go. We were all happy to have another brother or sister and all helped in getting things for the baby. My mother was busy knitting.

One Wednesday night in March, after we had all gone to bed, Mam was taken into hospital. We found out next morning and spent hours ringing to see if the baby had come. At 2pm we got the thrilling news that Mam had had a baby girl. We were all overjoyed. While Mam was in hospital, I had to stay with my aunty and cousins, Michele stayed with my Gran and Carl, Marylyn and Belinda stayed at home. After two weeks of anxious waiting we welcomed our mother home. I can picture her in my mind walking through the door with the baby whom she called Sian. She was so tiny and sweet and my mother was so pleased.

Now we were six kids. My father praised Carl as he was the only son and in a way he was spoilt. His name was on all the tractors and ladders and everything my father owned. Dad painted them all his favourite colour yellow. On his JCB it said 'H. C. Minett and Son'. Us girls spoilt him too.

I was only a child myself, but having the baby at home was fascinating and thrilling. I can remember how tiny Sian was, how entranced we were as she learned to smile. She had dark brown hair and big brown eyes. What a pretty baby she was. Life was good.

Chapter Two

21 October 1966

THE HOUSES OF Aberfan line the basin and lower edges of the Merthyr Valley, four miles south of the town of Merthyr Tydfil itself. The River Taff runs through the village and then 20 miles on to the south and the city of Cardiff. In 1966, Aberfan was a coal-mining community of about 4,000 people which, for generations, had lived with the buildings, equipment and structures of the south Wales coalfield, and the local Merthyr Vale colliery in particular. The apparatus of the coal industry had become so much a part of the landscape that they seemed to have always been a part of it, almost like natural features. These included the coal tips, the hills of waste which had been piled high around us. Before 21 October 1966, there was nothing to distinguish Aberfan from scores of other communities in the south Wales coalfield.

There were obviously dangers associated with coal mining. Memories of pit disasters such as that at Senghenydd in 1913, in which 439 people were killed, are in the DNA of those who live in the coalfield. But this was our home, this was where we lived. The miners who worked the collieries that created both the employment and the danger were among us; they were us. The slag heaps that darkened the countryside were a side product of the mines. Some were piled dangerously high. One had a spring underneath it. What effect could that be having?

We didn't know, of course. On that fateful Friday we all got up and dressed for school in the usual way. It was a very misty, dark, damp morning with fine drizzle. Rain had been falling most of the month. We could see that. But what we could not see was the destabilising effect that spring water was having under one of the waste tips.

Families carried on about their business unaware that the mundane rituals of everyday life were soon to have a terrible significance: the last time he tied his school laces, the last time she shouted 'Goodbye', the toy left in the centre of the rug...

As was often the case my brother, Carl, did not want to go to school but my mother told him he had to go. After all, it was the last day of term. Tomorrow, he could play.

We walked to school with our friends, all visiting the little tuck shop on the embankment on the way. We went in there every morning.

The front of Pantglas Junior School overlooked Moy Road. The classes at the back looked directly towards the mountains and the tips. To the right side of the junior school was the senior school and its playground.

We went inside and took off our coats.

On Fridays we had assembly and my memory is that that morning we sang 'All Things Bright and Beautiful' and the assembly went the normal way. Then we all went back to our classes and sat down. Mr Davies, our teacher, got out the board and wrote our maths class work.

I later wrote in my childhood journal: 'We were all working, and then it began. It was a tremendous rumbling sound and everybody in the school went silent. Everyone was petrified, afraid to move; we were frozen in our seats.'

The sound got louder and nearer. I knew something was going to hit us. Instinctively, I got up from my desk and tried to run for the door. I turned to see the black coming through the window. I was sort of dazed. The first thing that I felt was blood trickling down my face, and I thought my leg had gone as I couldn't feel it or see it.

I was stuck fast in the corner of the room with my hand crushed against the wall and I couldn't move it. I was in agonising pain.

Children were screaming and some were lucky to escape. My friend, Dawn Andrews, limped through the roof just as it collapsed. I can remember shouting to her to get help. I didn't cry as I didn't know what was happening. It was all like a dream. Bodies lay crushed and buried and the survivors lay looking at their best friends, dead. I could just see someone's hand through the crack in the wall. I didn't know whether it was a boy or a girl. I squeezed the hand and pinched it to see if it was alive but I could sense that the child was dead.

Next to me was David Bates. He had cut his head open and his face was blood red. I picked up a book; it was called *Through the Garden Gate*. That, too, was drenched in blood. I was too dazed and shocked to scream or to do anything but lay there reading this book, while others screamed for help. Then everyone started to panic as it seemed no help had come.

The muck was at least a yard and a half up the wall, the windows were smashed and nothing was left whole. The radiator, which was double thickness and heavy, had come off the wall and landed in front of me, covering my leg. This had saved my life as it had formed a barrier which stopped more muck from getting to my face and suffocating me.

I gave a sigh of relief as a man's face looked at us through the window. I guess what he was thinking as he looked at the mess in front of him was, 'Where do I start?'

Gerald Kirwan lay just beneath me. He, too, could not move as the muck was all around him, and there were desks and chairs piled high. Gerald lay next to another of our friends, who seemed to be asleep. We wondered if he was dead, but how could he be when there was nothing on top of him? Gerald later recalled his best friend who normally sat by him did not go to school that day. The friend that lay there sleeping had moved desks to take his place. Sadly he died.

Janett Hibbard was running on top of the muck, desperately scrabbling to get out, to get help. There were so many of us trapped and it took a long time to rescue us. A man came to our classroom. He could not open the door as it was sealed tight with muck reaching halfway up the door. Somehow the window that led into the hall had stayed whole. This man shouted for us to hide our faces as he smashed through the glass to get at us. Everyone was crying and screaming.

The man hauled dead bodies through the window, and then some of my friends who were cut. I just lay watching them.

Then, an amazing sight for me: I could see my Grandpa in the window. He looked around with tears in his eyes. I called out to him and he saw me. I wrote in my journal: 'It was then that I started to cry as I reached out my arms to him, but he couldn't come for me as there were too many bodies in the way.' It was a terrible moment – the feeling of waiting for someone to rescue you but knowing you were just out of their reach.

*

For many years, too many to remember, I have been compelled to find out exactly what happened to me that day, but also to know what impact the disaster had on my parents. I felt so close to them that I would not have the whole truth until I had it from their perspective. It is in my blood, I just cannot get away from those feelings and thoughts; they represented a kind a closure, that's the best word I can find for what I am searching for.

My parents' perspective meant I would not only get their response to what they went through, I'd also get their view of me. After all, I was only eight at the time. My parents could answer so many questions that I could not. What behaviours did I display? How did my personality change? How did I act in the days, weeks, years, afterwards? And what did all of this mean – and still mean – to them?

I felt a compulsion not only to listen to them but to tell them all, too. Maybe it was my way of finally closing the door on the past and letting my mam and dad know how I felt. Any survivor will know just how I feel. Years have gone by and I have always said to myself there will be tomorrow and that's when I will ask – but tomorrow never comes.

When my dad was very ill in 2014 and he came close to death, it was then I realised that if he died I might never know the truth. I said to myself then, that, when Dad gets better, I have to find the courage to ask him, otherwise, I will live with this guilt till the day I die, and I may never find peace within me.

During the months of Dad's rehabilitation, there was never a right moment to ask; I just couldn't prepare for that moment, and I agonised over how it might unfold if I tried to tackle it. I had for weeks prepared a script of questions that I wanted to ask Mam and Dad. My questions were deep, sensitive and daring. I needed to have the answers. For once in my life, I felt selfish really, with no real thought of how they would react and of what memories might be brought back to the front of their minds. I just needed to draw a line under the past once and for all.

When I finally took the plunge I chose a most inauspicious moment.

The busy family festivities of Christmas were over and the new year, 2015, had just began. I was sitting with my parents watching *Under the Hammer*, my dad's favourite programme. They happened to be auctioning a chest-of-drawers. Something clicked in my head and my heart was suddenly thumping as I blurted out, 'Remember, Dad, when me and Ben [my son] were hammering and breaking up that chest-of-drawers in the garden a few weeks ago? It was in pieces on the floor and then you said out of the blue, "That chest of drawers can tell a story!"'

Standing in the garden my father had then briefly told me that he and his friend Gwynfor Chamberlain, who was known to everyone as 'Ginno', went to the Pantglas Junior School a week after the disaster in his little yellow pick-up truck. They carried two chest-of-drawers from the ruins of the school, one from the hall and one from a classroom. This was *one of them*. When he said this I felt sick and panicked, to think I had destroyed it. I took pieces of the wood into the house to have a plaque made for the 50th anniversary. I'd immediately felt a massive relief to find out that Dad still had the other one and it was safe and intact in his shed. I just couldn't believe that only now, all these years on, I have a piece of history which bares the scars of that day in my possession. I wondered what was it that the chest had been used for? Had the little children kept their books in it? It freaked me out a little.

My mother was only 30 in 1966. I just can't take this in. She was the same age that one of my sons is now. What a life sentence to live through. Suddenly, it puts life into perspective.

Mam began to tell me her memories of the day. 'It was everywhere, everyone knew,' she said. 'I heard the news that day when Mr Jones the shop came running up the street shouting, "Iris! Iris! There is trouble down the school – quick, quick get there. Cliff has gone down."'

My mother carried my six-month-old sister, Sian, and gave her to Mrs Davies. Gran Minett came running up the street and together they all rushed down the hill. 'There was silence,' Mam said.

Grandpa Stan, my mother's father, had been on the bin lorry that morning down Moy Road with friends, Jack and Ken. They had been the first on the scene.

It was Grandpa Stan's face I saw as I lay trapped in the rubble.

It had taken Grandpa at least 15 minutes to get to me and then he couldn't move me. Jack and Ken got into the classroom to help but my hand was pressed tight against a wall and my leg had been broken. Whichever way I moved my hand, it made things worse. Eventually after a lot of tears and anxiety, they managed to get it free. The radiator, desks and rubble around me were then lifted, and I could see my leg. I looked down at it but it felt as though it didn't belong to me. I could never remember who picked me up because the pain was too great to bear, but whoever it was carried me to a window in the wall between our classroom and the hallway. I was lifted and passed through to someone else.

As I looked back I saw my friends. So many dead – and that look over my shoulder would be the last time for me to see them.

I was carried down the hall and through the door. Dead bodies lay on the floor and some were bloodstained. It was horrible to see. Outside, there were screams and shouts, and muck everywhere. Ambulance sirens were screeching and many of them were full of children. Mothers and fathers were standing in the muck looking for their children, their eyes searching through thick tears.

My leg was in such pain as I was given to my Gran Minett. How she managed to carry me I don't know, but she waded through the hell, over the debris, and through a monstrous nightmare, never to be forgotten. I can still picture my mother and father, my mother crying when she saw me. My dad carried me to the ambulance. I later wrote in my secret journal: 'My leg was in agony. It seemed to swing loose as if it did not belong to me.'

When Mam got to the school she said she saw my dad carrying me through the mud and slurry towards her. I had finally been dragged from the wreckage of the school.

I had no shoes on and I was crying for losing my shoes. She shouted: 'To hell with your bloody shoes! I will buy you loads of shoes!'

It was then I was passed to Gran Minett and taken straight to hospital in an ambulance. Gran came with me.

I was totally shocked, silent, on hearing this description of events. I had always thought that only Gran had carried me, not Dad. This means something special to me – to know I was in his arms at the school.

Mam remembers seeing her mother Mary and her brother Colin. They had difficulty getting through to the scene. Gran was crying, and shouting, 'My daughter's got three children in that school, let me through please, please!'

Mam waded through the slurry and the chaos as the sounds of sirens filled the air. She somehow managed to climb up into the nursery yard. By this time Dad had gone off to help out again.

Mrs Leyshon joined my mother. Her only daughter, Sandra, was missing. She was my best friend.

Of all the things, Mam wanted to pee and remembered going into the little toilet in the junior school which was still standing as it had not been touched. She remembered the silence again, and she sighed as she spoke of it. What must have been going through her mind God only knows.

Mrs Leyshon shouted for someone to look after my mother and then went off to look for Sandra. Mary and Gran Minett had to force my mother to go back home. There was nothing they could do, but wait.

I remember my ride in the ambulance well. I was laid on the stretcher as it raced through the roads with its siren blaring. My leg was in agony as the vehicle turned and tilted but we were quickly at St Tydfil's Hospital, a few miles up the road in Merthyr. A doctor carried me into a massive room where many of my friends were laying on beds. A nurse took my hand and told me I would be all right. Another doctor came and looked at my leg and put a splint on it. My hand had swollen up and the ring on my finger had to be cut off. This was the ring my grandfather had bought me only weeks before.

My head was bandaged but didn't need any stitches. I was given an injection for the pain. I was still feeling dazed and did not know what had happened or where I was. Nurses and doctors were racing around, seeing to the other patients. I was taken on a trolley to a lift and then to a ward which was called St Andrew's. I felt giddy and frightened. I was put on a high bed with a cover over my leg for the blankets to rest on it. Many children stared at me and asked me questions but I could not answer as I didn't know. I was in this ward cut off from what was happening outside and I got very worried.

An hour or more went by and I started screaming for my mother but it was no use. A nurse came to me with a tray and this frightened me. She told me to turn on my side for an injection in the top of my leg, but I wasn't willing and I screamed. She told another nurse to hold me and she stuck it in. My leg throbbed a little but I was all right.

I lay thinking, my young mind racing. It was hard for us to even work out what had happened. The first thoughts that came to me were of my brother and sister. I cried and a doctor hugged me and told me not to worry.

I felt terribly alone but my family was on its way.

*

On returning home, Mam made sure Sian was safe with Mrs Davies, and then – with her mother – she started to walk over to Merthyr Vale. They eventually flagged down a car to take them to St Tydfil's Hospital. The memory of that journey is raw for Mam. She arrived and asked a policeman to phone all the hospitals – East Glamorgan, Church Village, Maerdy, all over, anywhere – to see if Carl or Marylyn had been found. When their names did not come up on the lists of the rescued there must have been such a feeling of despair. On the ward, I was trying to close my tear-filled eyes when, to my delight, my mother and gran came in. My mother hugged me and wept on my shoulder. My gran hugged me too. My mam had to leave the room weeping and my gran comforted me. My mother was given a cup of tea and afterwards she came back in. I cried and asked her if Carl and Marylyn were all right? She said there was no news. They had not been found yet. I did not know what to do – cry, or say anything – but I just kept quiet for a minute and hoped it was not the last time for me to see them. My mother had to leave as the doctors had to see me. I said goodbye and she kissed me and said, 'I will see you tonight'.

Talking through all of this made my heart cry out in agony for my mother, but her and Dad seemed composed. They were holding their emotions together. Listening to Mam, there were no words I could say. Having children of my own and knowing the protective feelings that overwhelm you as a parent, I just wondered how Mam and Dad coped?

There was a long silence and then Mam said: 'I thought, well, I've got you and the girls.' Pause. 'We had to cope. For the rest of you. Somehow.'

Dad added: 'That's what pulled us through. We had to keep going for you. We just didn't have time to think.'

*

David Evans, known to all as Dai the Dyn (because he owned the Dynevor Arms in Troedyrhiw), was the man who made the first 999 call to Merthyr fire station that morning.

I met with Dai and he told me: 'We were one of the few houses with a phone in those days. A neighbour came running. He was quite upset and said there had been an accident in Moy Road. I asked what kind, and he said a house had collapsed. "Is it a gas explosion?" I said. "I really don't know," he said. I dialled 999, got through to Merthyr fire station and I told them I was speaking on behalf of a neighbour in Aberfan and that he had asked me to make the call to say that a house had collapsed on Moy Road. The fireman asked me to enlarge on that. While we were conversing people were running along the road and it was obvious there was a state of panic. So I said: "Hang on the phone. People are running past the house." I said: "This must be very, very bad." I remember one woman running here and saying: "The school has gone, the school has gone." Anyway, men started running, saying there were houses down, so I told the fire service, who were still on the phone, that houses had gone and the school had gone. And he said OK. They had not had any previous calls so I know I must have made the first call.'

Dai and the neighbour ran out into the street and to the Mackintosh Hotel. 'When I saw what happened from the Mac, what struck me was everything was so quiet, so quiet. All I could see was the apex of the roofs, piled up rubble spread on the roads. There was no road up to the school. And other people started coming and asking what the hell has gone wrong here and then pandemonium broke loose. The only way up to the school was a lane we called The Brook. I was met by smoke, people, mayhem, panic, steam coming from the rubble. And it was obvious it was a major disaster.'

Dai was 26 at the time and shocked by the destruction that confronted him. 'The thing that amazed me when I went up to Pantglas Road was the fact that the roofs of houses were intact and the exterior walls of the houses had crumbled and strewn across the road, so that the apex of the roof was on the road, intact and sitting on top of this rubble. Then the water started coming through the properties as the trunk of the water pipes had burst. It fractured and the water came rushing down through the houses, like liquid cement. It was thick, black. I remember in the houses of Moy Road the strength of the water had taken the windows and doors out.'

I said: 'You have lived here all your life. Do you sense that people want to talk?'

'Many seniors have passed on but there are many parents and siblings still here. It is still very raw for some, even your parents. I think it is still an open wound and they don't like to talk about it. Every year the memorial is held, I have purposely not attended because I think it is a very private affair. In my opinion, it's private for those directly concerned. You can't tell people to stop going, but I chose not to.'

Dai added: 'After that day the word Aberfan came in the English language to mean "tragedy, disaster". Even people abroad know it for that.'

Dai's father had been deputy head of the school. He knew all the pupils and staff, and had retired due to ill-health that March, only a few months before the disaster. The teacher who took his place was David Beynon, one of those who perished.

'My father was understandably very, very upset for his loss of colleagues and the loss of his children. He went into a shell for months. He would mention it all the time in conversation. He would bring it up all the time. Mrs Jennings, the headmistress who died, was coming to tea that day, she was coming to see my mother and father, and she had planned to come down the house on the Friday around lunch-time. She was killed. Yes, gosh, that day was a horror film, a horror film. There are no words that you can say for those who lost children. The courage of the bereaved was, oh my God alive, there are no words. I am glad I wasn't tested.'

It had been a stroke which had forced Dai's father to retire. The children had heard about his illness and wanted to wish him well.

'Only a few years back I was clearing out my father's belongings as I had kept them for a long time,' Dai explained. 'I flipped through papers and there was a "get well" card. It was from his class that he taught in 1966. The card was signed by numerous children. It broke my heart to see.'

I sighed, thinking of that remarkable card, from a group of schoolchildren who would so soon after be hit by terrible tragedy.

Dai said: 'I will have a look and see if your sister and brother's signatures are there.'

Chapter Three

Digging Against Time

THE 360,000 TONS of colliery waste known as tip No. 7 had been stacked on a hill 500 feet above the village. It had been becoming more and more unstable over the previous weeks, as the rain lashed down. Some of the material being piled on top was of a very fine consistency and, when it filled with water, it became something like a ready mix concrete. A deadly sludge carried on a thin sheet of water.

This moving mountain quickly reached a speed of between 30 and 40 miles an hour, crossed and filled the disused channel of the Glamorganshire Canal and ploughed through an old railway embankment on the outskirts of the village, just behind our school. A tip gang had been at the top of the mountain when the slide began. One of them later stated: 'I never expected it would cross the embankment behind the village which I could not see because of the mist which covered the whole of the village. There was nothing I could do. We had no telephone to give an alarm or any warning device. I shouted, but it was no good.' In the event, a phone call would have been no good either: the slurry was moving too fast.

It ploughed through more than half of the junior school, part of the senior school playground, through houses on one side of Moy Road and then the other. Its deathly grip reached deep into the village. It crushed cars and brick walls; it filled rooms with a thick, suffocating sludge. A car horn blared until the battery went dead.

In the hours after the disaster there was a frenzy of activity at the destroyed school as the search for my classmates, brother and sister continued. Eighteen houses had also been destroyed, as had a farm cottage which had been in the path of the slide. Human chains were formed to try to remove rubble.

Local miners took over much of the hard digging. Many had just come off shift; their lamps on their helmets were still blazing. One said later that they arrived to find the women clawing at the filth. 'Some had no skin left on their hands,' he recalled. 'Miners are a tough breed, we don't show our feelings, but some of the lads broke down.' The fire service broke windows to get into some of the wrecked classrooms. One fireman told television crews that the children he had helped rescue looked 'ghastly'. 'They were given morphine by the doctor and taken away,' he explained.

At 10.30am a newsflash broadcast news of the disaster to the nation. Very quickly people from all over were throwing a spade into their cars and heading to the scene. Everyone wanted to help. The rescuers were not allowed to use machinery in case there were people alive in the debris; they filled the lorries and drams with bare hands. Their fingers bled. Bulldozers only worked away from the main collapse site where rubble had cascaded down the road.

In one news bulletin, a reporter with a microphone stood on mounds of bricks in Moy Road as dozens of people moved wood and bricks with their bare hands. 'Every now and again everybody stops working, the bulldozers stop, and everybody listens to see if they can hear anybody under this rubble,' he said. 'I must admit standing here and looking at the wreckage it seems almost a hopeless task.'

Then he talked to a man climbing through the chaos. 'Do you think there is anybody still here?' he asked.

'My mother's probably in here,' replied the man before wandering further into the ruins.

Another man, whose face was black with coal dust, said: 'Well, I know where my son is at the moment, he is buried in that end classroom up there.'

Everybody in the village gave everything they could to help in the rescue. One resident told a local newspaper: 'We cut up cotton sheets for bandages, and gave blankets and pillows for the children as they were brought out on stretchers. Rescuers came in for everything, and we gave all we could. All we thought of was that children's lives were at stake. Everything lost its value in comparison with those children.'

The tip itself was still moving, thick sludge rolling through the street. Every now and again the digging stopped to listen for the sounds of someone trapped. But after 11am nobody was brought out alive. They were then bringing only bodies out. Each was on a stretcher, covered by a blanket. They were taken to Bethania Chapel and laid out on the seats. Their families came to identify them. A place of worship was now a makeshift morgue.

On the evening news a tearful Cliff Michelmore reported: 'Never in my life have I seen anything like this. I hope I shall never see anything like it again. For years, of course, the miners have been used to disaster. Today for the first time in history the roll call was called in the street. It was the miners' children.'

Grandpa Stan came home much later after rescuing me. He had been digging down at the wreck of the school for hours. He was black and blue with bruises from head to toe; his eyes were red raw and black with soot. Mam tried to stop him going back down by saying, 'You are not going anywhere, Dad', but she said: 'He just wouldn't listen. He just went, totally exhausted.'

The rescuers wondered if there might be people trapped in air pockets. They could not stop digging. Within 24 hours of the disaster, 100 bodies had been found.

A report into the disaster by consultant psychiatrist Dr J.M. Cuthill would later record that 52 boys and 52 girls died in the junior school, and six adults.

Six children died in the playground of the neighbouring senior school.

Six children and 22 adults died in houses. The majority of the dead had been asphyxiated.

*

'I know you,' said the total stranger. 'You are Cliff Minett's daughter. I saw you trapped underneath the rubble in your classroom with wooden beams over you.'

These words sent cold shivers down my spine. It was such an unexpected meeting in the most inauspicious of places. And it was to change my recollections of that day in 1966.

It was a Friday lunch-time in August and the sun was shining. I was working for an organisation helping people back to work. The manager, Tina Cook, and myself generally ate our lunches at our desks but as it was such a nice day we thought we'd get out. We walked through Merthyr and decided to get something to eat at a Wetherspoon's pub. It was already packed: some people seemed to be starting their Friday night very early!

Tina and I sat in the small restaurant with people packed on the tables at both sides. On one side there was a group of older gentlemen, all dressed in shirts and ties. They seemed to be very vocal and were having a good time. One of them, a tall man with a broad physique, suddenly approached our table. He looked very friendly but I couldn't help but think, 'Oh, here we go, I can't go anywhere without attracting someone!'

As he stood by us I realised that he seemed a little agitated. It was then that he spoke those words and I was totally taken aback. I had been completely unprepared to hear this here, now.

The way he said it showed that he was uncomfortable too. He had seemed to take a deep breath before just hurrying the words out; as if he was afraid he might not say them if he hesitated.

After he spoke I could feel a tingle from my feet through my body. I just didn't know what to say. I was in disbelief, but I was also very touched by the way he had approached me.

He went on: 'I had seen your father in the next classroom digging. I didn't tell him I saw you.'

This man just stopped me in my tracks. I just said, 'Please, please, don't tell me any more.'

I couldn't talk about it until I was prepared. Here was a witness to the most dreadful moment of my life and I had never known he existed until almost 50 years later. I had to think... I had to be prepared for what he could tell me. We exchanged our phone numbers and that's when I found out his name was Douglas Hughes and he was known to everyone as Dougie.

We went our separate ways, the sounds of the pub all around us, but my mind was elsewhere now. I had to meet him again, talk to him properly. He had now become part of my journey, a chapter I never knew would exist when I started out. My trip of self-discovery had been sent on a new course by this twist of fate, this accidental meeting.

It was a while before we met again. I was right to prepare. Dougie was able to reveal to me new memories. It had seemed to me that shock was no longer possible; I was wrong.

*

Dad became very ill in the months after I first met Dougie. My journey into the past took a sideline while I concentrated on supporting him and Mam.

But when I contacted Dougie again he agreed immediately to meet up and we arranged to see each other at the same Wetherspoon's pub.

As I approached I could see him waiting for me outside. He seemed a little nervous. I kissed him on the cheek and thanked him for meeting me. We walked in together and he insisted he bought me a drink. I took a shandy. We sat on the high stools right by the window with the warmth of the sun shining through. The place was busy, noisy, not the ideal place for me to record our chat, but I took the chance.

Dougie seemed more relaxed now the ice was broken and I was both relieved that we could meet again but also unprepared for what he might reveal.

He began to tell me his story. He was a former pupil of Pantglas Junior School and the neighbouring senior school. His parents, Doris and Berty, lived in Bryntaf. When the disaster struck he had been an apprentice miner at Merthyr Vale colliery, next to Aberfan.

'I was 19 at the time, working on the surface of the pit,' he said. 'Me and the colliery workers used to have a cup of tea and toast at half past eight with the workers in the fitting shop. After our break we were on our way to the washery room. Enos Sims was there and Brian Eddie and someone else I can't remember. Someone shouted over to us: "Enos, the tip has come down on the school." Enos was in charge. "Right," he said, "get in the car." Out of the colliery we rushed, over the bridge by the library. All this water was pouring down the hill and we thought what the hell is this? So we got up onto Aberfan Road. I got out of the car, walked up the steps which were clear, through the gates to the left of the school, and the first thing I saw has stuck in my mind ever since: a photographer was taking a picture of Susan Maybanks, the girl being carried in the arms of a policeman. I said to the photographer, "Stop taking photographs and give us a hand to help." At that time I didn't know that particular photograph, taken by Mel Parry, would go all around the world.'

Dougie had arrived at the very moment one of the most iconic photographs of the disaster was being snapped. Mel Parry was an apprentice newspaper photographer in Merthyr Tydfil at the time. Mel later said he got no pleasure from the success of the photograph as he wished he had never had had to take it.

Susan was my friend of course and Dougie had also known

her family. He knew mine too and was good friends with a neighbour of my parents.

Dougie went on: 'You had all the mist down that morning, and the trail of slurry was pouring down to the Mackintosh Hotel. The front of the school was standing and not so damaged. I just followed all the other rescuers, they all seemed to go to the right of the school where the windows were to the classrooms. They were high, I remember. I then climbed over the debris into the classrooms. I went into the school the front way which was still standing. I didn't realise how bad it was. I went into the classroom in the front, then to the left side, that was when I looked through the window which was smashed in. I could see you, Gaynor, I knew you. I saw you in the classroom. There was a few of you huddled together.'

I had a lump in my throat. I could not say anything. Just silence. I felt Dougie's discomfort at the memory. He was very emotional. The people and the noise around us in the pub had gone away now; there was just us. Just Dougie's words reforming this picture in our minds.

'I remember your leg was underneath you and there was a big beam on your legs. You had children around you. I didn't know how long they took to get it off your legs because I was out by then. You weren't crying, no. You were quiet, you were in shock. I then went into another classroom and I saw your father. Your father was digging, he didn't have a shovel. He was in the classroom, shovelling by hand all the debris. I didn't say anything to your father just in case, do you know what I mean? He would have gone in and made the situation worse. It would have only panicked him more, and then you more, so I kept silent, so I decided on impulse not to say anything.'

Dougie swallowed hard. There was a lot of pain in the way he spoke. 'Did you regret not telling my father, Doug?' I asked gently.

'No, I don't think so,' he said. 'Because if he'd known his child was in there it could have made the situation worse for you and him. If your father had gone in to where you were trapped, it would have got out of control; the men seeing to you would have been in control.' He shook his head. 'I always remember that bloody big beam across your legs...'

I told Dougie about what I had learnt about how my grandfather had come into the wrecked classroom to rescue me. He was as captivated as I had been moments before.

At one point he said that he never spoke about that day. 'It was the worst day of my life and that is it,' he said. He was telling me things he did not share with other people.

I asked him to go on with his memories.

'There were big steel props. I was trying to help the colliery miners put them in place to hold the school up. My hands were bleeding, my back was in half, I kept going.'

I asked him if at that early point in the rescue he and the others realised that so many had died.

'After seeing all the muck, I didn't think anyone could come out of there alive after 12pm. I couldn't imagine anybody surviving ten minutes let alone ten hours. I recall you lost your brother and sister that day too.'

Eventually Dougie was sent back to the colliery. 'The older men then took over from me. I was only young, I didn't know what to do. Then when the colliers came up, they came with their shovels...' He could not face going home. I asked him whether he thought what he had witnessed as a 19 year old had affected his life.

'No, I don't think so,' he said, but his mind was thinking of someone he had known in Aberfan...

'I am going to tell you a story now, Gaynor. On the Thursday night, the night before the Aberfan disaster, he and I went down to the Navigation pub, down by the steps going up to Perth-y-gleision. I remember it had been raining for weeks on end. He said to me, "Doug, have you got any matches for my cigarette?" I gave him the matches and shouted to him, "Watch you don't kill yourself!" He was gone the following day.' Dougie looks uncomfortable. 'I haven't said that to anyone since, you know what I mean?'

I said: 'So consciously you are still affected by that day, Doug. You may not realise it, but you are.'

I asked him what it was like in the colliery after the disaster. 'Desperately sad. They were all sad, the atmosphere was awful.'

Did he think the miners felt guilty for tipping the coal?

Dougie does not answer directly. Instead, he said: 'I am going back to when I was younger now, when I was 11. I used to collect coal on the tips, all us kids we all did the same. There was a stream coming down then. We used to bank the water up, put the coal, anything we could use, and bank it up so the water would form a pool and we would swim in it and have fun.'

I remembered doing that too on the other side of the mountain with Ronald, Gerald and Paul. This was the water from the underground spring. No-one took much notice of it.

Dougie said: 'The miners, they must have felt guilty for tipping over the stream. Anyone could see that stream, even me as a child and then as an adult. When they started tipping on tip No. 7, there was a stream there, I know that for sure. I can tell you now I saw it.'

Dougie said he did not go to the inquiry into the disaster but heard they had used 'jargon' when talking about the springs. I said they decided that geographical factors had caused accidental death in every case; that it was not manslaughter. He raised his voice as he replied: 'No! No! No! It was man-made. We had to go up the tip often after the disaster, to monitor the water coming down from the mountain, me and Enos Sims. I remember we had to use a V Notch Weir, which we placed in the stream: it obstructs the flow of water causing the water to flow over the v notch which measures the flow, pressure, quality of the water coming down from the tips. The water then comes into the tank with measurements which we monitored. Following those observations, the coal board then had a culvert made.'

I had heard this. In fact, there is still water coming out of there into the pipe system and drainage even today, you can hear it running.

Dougie said: 'There are still small mines up there, at different levels. The brown in the stream is the iron materials from the levels.'

I take a moment to think: after it all, the death, the destruction, the spring keeps flowing. Something in nature does not stop.

Dougie started going back to the day. In all my conversations on this journey this happened: whatever we went on to talk about we were drawn back to that single day, as if we were anchored to it in some way.

'Everything about that day has stuck with me for years,' he said. 'The main thing was the people who just got stuck in and helped. Because you were stuck in the school and trapped, more could have come down on you again, the tip could have moved again. I am thankful that you got out safe. I have never spoken about it because I felt that people would think I am boasting and everyone was there to help so I kept quiet. All of my friends, they all helped out.'

All, he said, were angry when the NCB took money from the disaster fund to remove the tips. 'I gave all my wages to the fund and then to find out that the Government took that back! Money which I had contributed along with the other miners who also gave their wages! We didn't earn much but we gave it all. This made me so angry and that's anger that is still current today.'

Dougie eventually left the colliery and settled in a house in Merthyr Vale with his wife and child. I had not thought about that part of his life, the life after the disaster, but he told me it was a part of my story too.

'Gaynor,' he said. 'I have one more surprise to tell you. My wife, Karolyn, nursed you in St Tydfil's, she was there that day.'

He reached into a bag and brought out a photograph of a young woman in her nurse's uniform. He told me this was

Karolyn, aged 20. 'She won an award for nurse of the year following Aberfan,' he said. 'She remembers you in hospital vividly.'

Dougie's wife had been a trainee nurse. She was single then, and called Karolyn Howells. She had not met Doug until New Year's Eve, 1966.

I took the photograph. 'Oh, my God, I am flabbergasted,' I said. I mentioned Karil, a nurse I had noted in my childhood diary, and that she had only recently passed away. 'I have to meet her, I can't wait.'

Dougie said: 'Ring her now, she is at home.'

I did. A soft voice answered. 'Karolyn,' I said. 'It's Gaynor. Doug is here with me. He has told me. Can I come and see you tomorrow?'

She said yes. I was so delighted. One piece of a jigsaw puzzle leads to another. I said goodbye to Doug knowing that he had given me so much; I only hoped that meeting me had helped him get over some of the anguish and frustration he had been carrying with him for almost half a century.

*

Karolyn loved her garden. It was there that we were to meet. It was a beautiful location to share memories.

The sun shone on us and trees and plants bordered the back garden of her and Dougie's home. There was a pile of freshly-picked gooseberries on the patio table.

Karolyn spoke very quietly, but I knew immediately we had a connection and a bond. Dougie remained very quiet, listening to Karolyn, as he too had not heard her recollections in full. Neither of us asked questions; we just listened to her story.

'I was only 20 at the time and I had never heard of Aberfan. I was from Aberdare and I was working at St Tydfil's in the orthopaedic ward. I was just going on my break at St Tydfil's, I had just changed my clothes – my uniform – as I had planned to pop down to town, and I was walking towards the entrance at the hospital gates and the porter told me to go back and get changed to my uniform. "Love," he said, "there is a fire in Aberfan school and they need all the nurses they can get. Quickly go and change!" I went back and changed and I was one of six nurses. Len Goodwin was there, he was one of the charge nurses on St James ward; one of our training tutors Marge was also there; we all went down [to Aberfan]. I just got out of the van and I was up to my knees in mud, it was just dreadful.'

The unpleasantness of the memory showed in Karolyn's face. She took a deep breath. 'I have never seen anything like this in all my nursing career. I was just looking and I... I just couldn't take it in at first... I said, "I thought there was a fire?" I wasn't prepared for this at all. I was in my first year of nurse training. We were all just astounded, it was so, so horrible, you know, Gaynor, devastating. I don't think anyone would ever be prepared for the sights that day. I didn't see many children coming out alive...'

None of us spoke in the pause that followed. Then Karolyn went on: 'We were down there at the school for hours. I think I came back about 7pm because we were starting to get... well, crying. I was just crying, but I never knew I was crying. Tears were streaming down on my face all day, in the end I just took no notice. Do you know what I mean, Gaynor? We just didn't realise we were crying, all of us. But I remember five children's bodies coming out...'

Something about the beautiful way she described the tears, the agony of the helpers, made me sigh out loud.

'I washed the children that perished,' she said. 'I was also drawing up morphine for the ones that were still... you know, what I mean? The medical tutor, Marge, was administering it, she was the Sister. We were sent back to the hospital later, as the day went on. We were all getting hysterical, that's the only word I can describe for it. Hysteria was setting in with all of us, we just couldn't handle it any more.'

I said that in the archive footage I have seen the parents always seemed so calm.

Karolyn said: 'They were just watching and waiting for news, waiting to see who was going to come out alive.'

She said she returned to the hospital 12 hours into her shift, but her day was not over. 'I showered, got into a clean uniform, then I went to A&E to work for a few hours while they were bringing the children in. Then up to the children's ward where I saw you. You had a fractured femur.'

'That's right.'

'I remember a few little boys there too. I got home about ten that evening.'

Dougie broke in to mention again how he had seen me in the classroom, how he had spoken all about it for the first time only the day before. 'I never talked about it before,' he said.

'I didn't either, we just got on with it really,' said Karolyn. 'My parents saw it on the news and, because I hadn't got home, they were really worried. They knew nurses had gone down there, and they were so worried when I got home. I remember I still had the mud in my hair, even though I had showered; it stuck like clay into my hair. One of the nurses finished straight after her experience, she couldn't go on. It ruined the start of her career. It seems for me that going onto the children's ward helped me, because we were nursing the children that were still alive, you know? It was a very long time before I spoke about it. I told my children about it when they got older but you push it back into your mind, because it was such a devastating thing. I was only young, I have never seen anything like it in my life, you know, the tip, the mud, the horror.'

'There is no word in the vocabulary that can describe it,' I said.

'No, you couldn't,' Karolyn replied. 'It was children that made it worse. Unbearable. First of all, I had awful nightmares. I couldn't sleep, I couldn't even talk to my parents about it. They would ask me, I just couldn't tell anyone... I can't remember how long it took me to tell them but it was a long time after.

My dad told me, "I can't understand how they could have sent you, knowing you were only training." I remember, though, that after my experience it made me stronger, made me more determined to finish my nursing career because something like that could happen again. We need nurses, doctors. But when I did tell my children when they were older, they were very shocked to know I was there.'

I asked Karolyn to clarify which ward I had been on when she was looking after me. 'The children's ward was St Andrew's; St Peter's was geriatric. St James was the male orthopaedic ward. St Andrew's, your ward, Gaynor, was above the old A&E.'

She looked at me as if she could see me all those years ago. 'Gaynor Minett,' she said. 'I remember you in the ward, I remembered the name. There were lots of other children, but yours I always remembered. I nursed you for a month. Because it wasn't my regular ward, when I left there, I just carried on with my duties. The ward was always busy, but cheerful; we were trying to keep the children cheerful, trying to keep parents' spirits up. You couldn't walk around with a miserable face.'

I showed Karolyn the photograph of me with the comedian Alan Taylor. 'I remember that photo,' she said. 'I tried to get away from Alan Taylor because I didn't want my photo taken. Alan was kissing all the nurses, and one of those little boys said, "Where are you sneaking off to, Nurse Howells?" The kids were so resilient. They just loved to joke with the nurses all the time. It wasn't a sad ward. I was trying to keep their spirits up but also their spirits kept us going too. You were quiet first of all, then you started to come out a bit, like the other children, I used to see you laughing. Knowing you had lost your brother and sister, you just keep the emotion inside of you. I remember you, I always have. I didn't see you after, I never knew how you got on.'

Another memory struck Karolyn. She had nursed one of the people of Aberfan on the orthopaedic ward before the disaster. He was only a young man but he had broken his leg. 'He had only been discharged and gone home the week before the disaster,' she said. 'I was later to find out that he died in one of the houses on Moy Road. If he had still been in hospital, he would have been still alive. The tip came down on his house.'

I said to Karolyn that I had often thought that I could have simply died instantly like so many others that day; that I would not have suffered, I just wouldn't have woken up.

Karolyn took a moment and then said: 'I look back now and I think it would have been better if I had talked about it first of all, but it just wasn't something you could talk about or bring up in conversation. I was watching it on the news and thought, "I couldn't have been there surely, it's not real, that couldn't have been me". I haven't talked about it for years until Dougie came home and told me he had told you. A lot of people do ask when it comes around, and I sometimes say I was actually down there when it happened. They are fascinated, the younger ones.'

I said: 'None of my family spoke after, and I didn't for fear of upsetting my parents. If I hadn't started writing my diary, perhaps my head wouldn't have coped with it as a kid. Many have died young, suffered severe mental health all because of that day.'

Karolyn said: 'My experience made me more protective and worried for my children. I lived in Merthyr Vale when I had children and I knew the school was safe, but it was always at the back of my mind: if it happens once it can happen again. It ruined people's lives. It split the village up. Every anniversary brings it all back, you can't help it. But it's today, Gaynor, that I've been remembering a lot of details that I have always pushed back in my mind.'

I turned to her husband. 'I am so glad that you came to me, Doug, to talk. It's putting the jigsaw pieces together.'

I could see it was not so clear for Doug. The pain of seeing me in the room and of not telling my father for fear of panicking him and me still haunted Doug. 'Should I have told your father?' he said. 'You were in that classroom trapped and that... bugs me. Would it have made a difference to you and your father?'

It seems like this is a question Dougie has spent 50 years trying to get right in his mind. I feel incredibly uneasy trying to answer it. I know if I can say the right words it will give Doug the closure he deserves.

'I have been thinking about that since we spoke, Doug,' I said. 'You seemed to be carrying that as guilt. It's not healthy. If I was a child laying there and you didn't know who I was, I think you would have felt guilty all your life, you would have always wondered what happened to that girl... I was someone you knew and you knew my family, you knew I was getting rescued.'

'I think I was right,' Dougie said. 'I made the right choice not to tell him.'

'Yes, Doug, you were right,' I said. 'I typed up your notes last night and I could sense a feeling of despair in your voice making that decision.'

I promised Dougie that I would tell my dad, allow him to put this ghost he is carrying to sleep.

'Yes, closure for you, Doug,' Karolyn said.

Then she turned to me and asked me about my dad.

'He's just come out of hospital,' I said.

'Which hospital?'

'Prince Charles,' I said. 'CDU.'

Karolyn lifted her hands. 'I work on CDU!'

'His name is Cliff Minett,' I said.

'I'm lost for words,' she said, 'I cared for him two weeks ago. I work nights. Fate, it has to be fate.'

I think our meeting exhausted us all. Dougie had worried so long over a split-second decision which, in the end, whichever way he had decided, would have made no difference. I had learned more about that day, been transformed back into the child in danger and the child in hospital; built up a greater picture of the nightmare. And Karolyn, a caring nurse, had been able to talk openly about a major trauma right at the outset of her career, which could have ended it but instead made her stronger.

I was proud of us three for talking openly, sharing our most private feelings. We had been brave to go back to that different place and time, face up to our reactions and our emotions. Along the journey I have always found myself being shocked by tiny revelations about the day of the disaster. Sometimes shocks come while you are sitting in the sunshine with two friends you hardly know, among the flowers of a beautiful garden.

*

When I asked my dad in detail about the rescue the thing he remembered was the feeling of excessive exhaustion. He had worked down the pits for 20 years and had 'never, ever' felt as disorientated. 'We worked flat out, flat out,' he told me.

He said the farm which had been in the path of the tip had ended up in the school yard. Chickens from the farm were flying up in the air through the roof: they were still in their pen. He remembers the slates flying off and the chickens flying out. He couldn't believe they were still alive.

He said he remembered my classroom. 'The children couldn't get out. You could only get out of the windows. The roof had collapsed. Some were breaking the glass. Certain classrooms were hit the most. The L-shape part of the school had all gone.'

Later my dad wanted to visit me. There were police everywhere around the village. Sergeant Ford stopped him. 'Where are you going? You need permission to leave the village and come back.' Dad shouted, 'My daughter is up there in hospital. Let me through!'

There were so many people walking over the mountains, people came from everywhere to help. Others were searching, like my dad.

He did not know how many children he had lost. But he knew where I was, and he was determined to see me.

Chapter Four

The Courage of the Vulnerable

THREE CLASSROOMS HAD completely disappeared under the thick slurry. Five teachers died in all, including our head, Miss Jennings, whose body was later found in her room. She had been a year from her retirement.

Mrs Margretta Bates's scholarship class, studying for their upcoming 11+ exams, had taken the full brunt of the huge wave of dirt. Margretta was 35. She had two children of her own.

Her colleague, David Beynon, tried to shield a group of nine year olds. He was found cradling a number of his pupils. They were dead in his arms. David left a wife and son.

Mrs Madge Rees was just 22. She had only recently been married.

The youngest teacher was Michael Davies. He was 21 and was in his first year in the profession. He was my teacher.

Gerald Kirwan, who lay next to me that day, later recalled being carried to a nearby house. He was in deep shock and said he could not walk because he felt so stiff. He told me that he dirtied himself and was deeply ashamed of anyone finding out. He was so embarrassed, as nine-year-old boys didn't dirty themselves, he thought. In the house, he stepped over a girl who lay still on the floor, cleaned himself up in the toilet before being taken to hospital.

Janett Hibbard, who climbed free to go for help, recalled that she always had a guilty feeling of leaving her friends behind and not returning to the classroom to try and rescue those who were trapped. Janett is still haunted by the memory.

Dawn Andrews, my friend, remembered running home to her mother and telling her that the school had fallen down.

Her mother gave her a row for lying and she had thought she had fallen in the muck because of the state of her dress.

My dear friend Susan Maybanks was almost completely buried but managed to stick her fingers through a gap in the muck and call to someone to save her. Only on seeing her finger poke through a hole did they know where to find her. The picture of her being carried out by a policeman was printed around the world in the newspapers.

Phillip Thomas remembered crying as the stones crushed his hands, and calling out for his mother. He had left school on an errand with another boy and was out of the classroom when the mud dashed down. Muddy water poured over him. He lost three fingers, and sustained an injured leg and fractured pelvis. His ear was also ripped off and had to be sewn back on. If the mud had not caked solid around him he would have bled to death from his injuries. The boy on the errand with him was found dead two days later.

Another pupil had left her classroom to take dinner money to the school office. When they eventually pulled her free, she was still gripping her shilling dinner money tightly in her hand. She still keeps her shilling as a memento today.

Just as on any day, there were children who did not go to school. I have often wondered how those who escaped that day felt. I have spoken with a friend, who was a couple of years older than me. He survived because he slept late that day. It meant the difference between life and certain death for him: he was in my sister's class. 'They all died except me who was late and another boy who had a day off,' he told me. 'When someone asks me, "Was I affected?", well, it is bound to affect us all. I was ten-and-a-half.' He has lived a life he might never have had.

Four teachers thankfully escaped with their lives that day. Pantglas was Hettie Taylor's first job. She was just 22. A few moments before the disaster she had been talking to a colleague about plans for the following day's staff party. Standing in her classroom, she thought that a plane was crashing on top of the school. She told her class to get under the desks as the walls began to crack. She opened the door and it was just black outside. She told the children it was a fire drill. She told them to not look right or left, just go straight home. Her pupils escaped. The school had always had regular fire drills and Miss Jennings had stressed to staff to always take the register with them. This she did now, grabbing it and handing it to the policeman outside. She stayed at the school all day to give the names of the children that were in the classes that were hit.

The teachers were all good friends. They were at various stages of their careers, and together they looked after the education of 240 pupils with love and kindness. Their training had taught them to be attentive and considerate teachers; nothing had prepared them for this.

*

With the deaths of so many schoolchildren, the suffering of the teachers has some times been overlooked.

They had been preparing for the last day of term and looking forward to the next day's staff night out.

Howell Williams was 22 at the time. He has been largely unable to talk about the day. Twenty years ago he contributed to one newspaper article. 'I got my £1 note ready in my hand to pay for tomorrow's staff night out. One of my colleagues, Marjorie Rees, was organising a night out for the staff. There was a feeling of happiness in the air. The school football team had won a match the night before and as I left the staff room a blond-haired boy charged past me. He shrieked, punching the air in victory. "We did it, sir, we won." I grinned back at him.'

He went to his classroom and called the register. The children set to work, writing about things they had done the previous day.

He looked out the window, knowing the familiar view to expect: the school yard and, beyond, the almost vertical hillside leading to the towering slurry heaps above. 'But this morning it was different, through the heavy fog that clung to the valley, I saw something moving fast, a huge boulder whirling, spinning down the hill side. Chasing behind it was line of blackness. I turned back to the class, opened my mouth to shout a warning but it was too late. The boulder crashed into the school and a sea of black coal and slurry hit our classroom. Like a suffocating porridge, sweeping everything and everyone with it. I don't remember the next few moments, only the sensation of being carried up and forward. When I came to a halt I was pinned against the wall at the front of the class, behind me was a towering mound of slurry that had slipped down the hillside.'

*

Howell answered the door and I said: 'How are you?'

He smiled: 'Here we are. Getting older. I am not so bad.'

'Time flies,' I said. 'I am 57 now; time goes by too quickly.'

As we started to talk it quickly felt like we had never lost touch.

I told him I had spoken to other teachers and they all sent their regards. 'I have lost touch with them, I haven't been well,' he said.

Howell had lost his wife. His grief was palpable.

'It was hard for my wife after the disaster,' he said. 'You lead separate lives when something happens to you; it's difficult for anyone to understand.'

The survivors have travelled this journey together, whether we have been side by side or not. Because of that common understanding we connect very easily in our conversations. Maybe this can shut other people out.

Howell asked me what I was doing now and I told him I had just got a job with Barnardo's. Howell, by now retired, said he liked to play golf, garden and watch TV. He had collected together a number of photographs for me to see.

'When I left Ynysowen Primary School at Aberfan I went to

Bishop Hedley and Goitre Comprehensive, secondary modern,' he said. 'I then did my degree, here is my photo [in his cap and gown]. My passion had always been to be a headmaster and I failed to achieve that because you have to have credentials locally. I remember teaching in room six in Goitre. I had tingles down my spine, and I then had a breakdown. The head brought me home. I didn't know then what it was. I was treated at the local hospital. I take tablets now to make me feel better.'

I said nobody quite understands, do they?

'Nobody,' he said. 'One day I went up the cemetery and do you know what my thoughts were? What the hell am I doing here?'

*

In his writings about that day, Howell had remembered the desks and chairs buried under the rubble in his classroom; and the silence which was only broken when the children began to come to their senses. 'Please, sir, help!'

'One girl kicked the door to get out. I don't know why but I told her off for damaging school property, it was as if I was trying to retain some feeling of normality by being a teacher. She turned around and darted out saying, "I am going home, sir". I injured my foot, I slipped off my shoe to free it. I tried to claw at the slurry.'

Every time he moved some of the slurry more fell in its place.

*

Howell brings a book of photographs over to where I am sat on the sofa.

Howell said: 'I spoke to you when the Queen came in 1997. What a memorable day that was. Remember we all went for a meal at the Castle Hotel in Merthyr? None of us had spoken about that day, until the Queen came. We shared our memories, feelings, for the very first time. It was an emotional day for us all. We all got drunk and laughed together.'

We were looking at the photograph, but Howell's mind was elsewhere.

'It was so powerful, the disaster, these eyes see... I actually saw the slide of the disaster coming down. Did you see it coming down?'

'Well, it came through the window, and I remember looking up and seeing the black coming through the window. I managed to get up from my desk and run for the door, but I never made it to the door.'

Howell replied: 'I was standing up. I had no control over what happened to me. I was pushed side to side. I can always remember when I took my class up onto the Canal Bank days before the disaster, and we were all collecting catkins. I always feel sorry for one boy, who used to do gymnastics in the hall. He died. I was standing up and he was in there... He died. What a life, eh? My desk was there, I gave out the new books and they were sitting there in my class. I survived it. There was a girl, she was trapped, they were all trapped. I said to them, "I will be back".'

*

Howell scrambled free from the school, the coal dust thick in his nostrils. 'I felt so useless I wanted to sink to the ground. I knew my pupils were waiting for me inside. A group of men stood opposite and said that help was coming. I climbed back into the school and I saw a frieze of witches the children had painted for Halloween, ghoulish faces peered out in the gloom. I went back to my classroom and started to claw at the slurry to rescue my pupils.'

*

The teachers I had met had raised a doubt in my mind about

one of my memories of the day. I wondered if Howell could clear it up for me.

'Did we have a service that morning in the hall and sing "All Things Bright and Beautiful"?'

'No,' he said.

'I can't understand. Why did we think that?'

Howell shrugged. 'Mrs Williams used to take the dinner money in the hall. I sent two pupils over from the class to the secondary school with the dinner money. I was in my class, it is as simple as that. If we had been in the hall for assembly, we would have all survived; all survived, yes.'

Oh, God. I'd always believed there had been an assembly, with us singing that beautiful hymn. There hadn't. If there had been the children might have been saved.

'There is a cross up there,' Howell said, referring to the cross-shaped panelled memorial in the cemetery which lists the names of the dead. 'I have been up there loads of times. I taught them all. When you're gone, you're gone, that is the thing. Only I know what it was really like in there in that room. A little girl trapped by a window, another boy was up on the ceiling, climbing down.'

I am overcome. 'Oh, bless you, Howell. You were highly respected as a teacher, and a good teacher, never forget that. It was very brave of you to go on teaching, Mr Williams. Teaching after must have been hard. You had determination and guts to go on; you should be proud of yourself.'

*

Howell wrote: 'Time stood still, minutes felt like hours. A voice said, "It is all right, rescuers are here". I saw a nurse, her face like an angel. "Thank God," I whispered. I was led from the school physically and emotionally drained. Hundreds stood outside. I was taken back home, told to rest. It was impossible. Switching on the TV, every report filled me with despair. Only I and three other teachers survived. I was later to find out that

Marjorie Rees was found still clutching the £1 note I had given her. I felt guilty that I survived, surely the children who died had more right to live than I.'

*

Howell shows me a photo of his wife and children. A family photo taken before the disaster.

Howell went through the stressful ordeal of giving evidence at the tribunal. 'There were all these solicitors and lawyers. I went down to give evidence with the other teachers and all the press outside, but they all rushed down to Aberfan that day because there was reports of another slide there.' What did you think of what happened in the end? 'They just got away with it all.'

I said: 'It is difficult, Howell. PTSD was not known then. I think we all have it, or had it in our own way.'

Howell: 'I finished teaching in 1990. I wanted to be this wonderful headmaster but it didn't work out...'

'Do you get out and about much now?'

'Not really.'

'I always have the memory of you teaching me at the prefabricated classrooms down the park which were makeshift classrooms after the disaster,' I said. 'The topic you were teaching then was India, how the tea was extracted and what the Indian women wore. We were being filmed by reporters and I remember saying, "It always rained in Aberfan, so we wouldn't be able to grow tea here". I've remembered that conversation all my life.'

*

Howell wrote: 'It would have been easy to give up in those long bleak days but I felt I owed it to the children to keep going. So when a makeshift school was set up in Aberfan I went back to teach. Getting back to normality was impossible. Then in 1968

71

the new school was opened in the village called Ynysowen School, but I knew that if I was ever going to come to terms with what had happened, I had to be free of Aberfan. Shortly after the official opening, a year later by Harold Wilson, I left to teach in Merthyr Tydfil.'

*

'Here is a picture of me in Ynysowen Primary School the day it opened,' Howell said. 'Here's a picture of me in the classroom with Mary Wilson, Harold Wilson's wife. Harold signed the visitors' book at the school. I remember he didn't sign it "10 Downing Street", only his signature.'

I said: 'I remember that well. I was chosen as the girl to present Mary with a bouquet of flowers, and I placed a rose into the jacket of Harold Wilson.' We both looked down at the photo. 'I am amazed that you have kept these, I am so surprised.'

'There is the mayor, all eyes on Mary, look.'

Howell led me over to the window sill. More photos of Aberfan, school photos. I can't believe all the memories they unlock for me.

'You went to the Isle of Man, didn't you?' he said. 'Do you remember this?'

'Oh, my God!' An old photo from a school trip. Me and my friends. I am wearing a short pink pinafore slip, white cotton blouse, and long socks to the knee with tan pump shoes. There are lovely yellow flowers behind us. 'I've never seen that.' The names of the smiling friends just roll off my tongue. 'Oh, Dawn Adams, my close friend there. I haven't seen her for many years. Susan Maybanks, Christine Jenkins and Janet Jones and me.'

This was after the disaster. 'I remember the Manx cats with no tails, the trams in the middle of the roads, we never saw any cars then. We were only about 11 or 12 in these photos.'

'Do you want them?' said Howell kindly.

'No, I won't take them, Howell, but I will take a copy of

them and bring them back. These are yours, you have treasured them for 50 years.'

Howell points to another one. 'This was the first year Ynysowen Junior School won the Buckland Challenge Cup, and the Robert Minney Cup.'

'Oh, look, there's my cousin, Terry Minett. He called at my house yesterday. Oh my God, all the faces and names I had forgotten.'

I held the pile of photos and flipped the top one to the bottom. It reveals another memory.

'Oh, God, there's me in Llangrannog, with David Davies, David James, Susan Maybanks, Diane Fudge...' I pointed to them all in the line and could name them all. Twenty of us as children, all of whom had survived.

'It rained all weekend,' said Howell of the trip. 'It poured down the Friday night when we got there.'

'I remember the donkey poking his head into my top bunk in our cabin, causing quite a fright! Oh Howell, I am so glad you kept them... They are so historic with so much meaning behind them.'

I turned over another photo to show Howell with the school football team and another teacher.

'This one is Mrs Jennings, do you remember her?'

'Yes,' I said.

'She lost her life.'

The football team were smiling, holding a cup, and sitting on long wooden benches. I could remember the benches from the school gym.

Howell said: 'The last words Mrs Jennings said to me was... I was on £37 a month then I think, and she said, "The cheques have arrived Mr Williams". I can remember these things you know. I know it's not important now but who would have thought someone like that would have died instantly?'

Howell sighed and then moved on: 'I also captained Quakers Yard Junior School, we never lost a cup match.'

I told him that he had achieved a lot with his life, playing rugby for four rugby clubs, managing football teams. 'You made good things happen for the younger generation of Aberfan and you inspired them at an awful time to go on and achieve their dreams and ambitions.'

'I was at the sharp end after the disaster,' he said. 'When I go, all that will have gone with me.'

We reminisced over many more old photographs. Howell shows me a photograph of himself when he was a baby. 'You're like your mother,' I told him.

It is lovely to see him laughing, enjoying some of the past instead of being so upset by it. 'You have a brilliant memory for names,' I told him.

There are photographs of his college days with his friends, family photos when he used to have a caravan by the sea with his family. We talk about his children, grandchildren and their ages and what they are doing now. I tell Howell about my children and family. We laugh at the clothes we wore in the 1960s. 'Every photograph is important to me,' he said.

Chapter Five

Angels in a Disaster

IN HOSPITAL, IN the hours after the disaster, I had had X-rays on my leg by a wonderful doctor called Gordon. Then they had moved me into a different ward with a higher bed, as they had to build bars around my bed and put my legs in traction. My leg was put on a pulley wheel with weights at the end of the bed. This was painful and I had to stay like this on my back for 12 weeks.

After they had settled me, I was soon feeling all right. I had one of my friends by the side of me. Her name was Susan Maybanks. She had sprained her ankle and she talked to me, keeping me company. We could tell what had happened as it was all on TV and the wireless.

It was horrible to see and hard to believe. The day went by with casualties pouring in, but just the one or two badly hurt. That night all the parents arrived. They looked pale and their eyes were red from weeping.

My mother and father sat down with my grandfather. My father talked to me. I felt so hurt to see them so helpless.

Then I asked my father if Carl and Marylyn were all right. He replied in a soft voice, 'They are gone. They went to heaven with the angels.'

I just couldn't believe it. I didn't cry but just hugged my father. I felt so weak, all the strength drained from me. I would never see them again. Carl was just seven; Marylyn was ten.

Then I was also told my best friends, Sandra Leyshon and Desmond Carpenter, and most of my other friends had gone too. Paul Davies, my classmate, who sat right next to me in class, had also died. When, as a young teen, I tried to confront this horror in my secret journal, I wrote: 'How can inches between us, as we sat in class, be a position of life and death?' For me it was a miracle, a miracle that meant life for me.

My parents were too upset to do anything. We all hugged each other, as there was nothing else we could we do.

Then I let my parents go home.

The following day, in the afternoon, I stayed awake while the rest were sleeping. I lay unable to sleep and dreamt of what had happened and suddenly I broke down. I sobbed and tears trickled down my face, tears of death. The doctor came to me and asked what was wrong. I told him and he hugged me tight and the expression on his face was so sad. I sobbed on his shoulder and I can still see his face. He lay me down and gave me a sedative and told me they were gone to Heaven and they would be loved there as God took care of them. I quietened down and shut my eyes. His hands went through my hair. His feeling for my sorrow was there in his heart too.

The morning came along with more newspaper reports and wireless news broadcasts. I listened but took not too much notice. We had our breakfast and the doctors made their rounds. My parents came to see me every day and they both looked full of sorrow.

The ward I was in was St Andrew's, which was about three floors up. My bed lay on the left hand side by the window. Outside the window was a huge chimney stack which frightened me. Like the coal waste, it was ugly and black. When my parents came to visit me, I would get upset and tell my mother the chimney was going to fall on me. My bed had to be moved to the other side of the ward where I couldn't see the chimney stack.

My parents came to see me every day, except one. October 27 was the day of the funerals. Eighty-one of the victims were buried that day. A generation of the village. All were children, except a mother who was buried with her two sons either side of her. Hearses lined the streets. The graveyard was not big enough.

My mother and father described being so numbed by the experience that, to this day, Mam doesn't remember who came to visit me that day. The hospital visits had put a huge strain on my parents. I never realised until now how intense this was for them. I was a child and only thought of how I was feeling. Mam said to me recently: 'You lost your brother and sister but we still had to come and see you and look after the family, the baby.' Dad added: 'I don't know how we did it, Iris.'

I had many cards and presents and lots of visitors. One present that I had was a doll and a pram from Princess Margaret. So many people tried to make us happy. So many people were so kind and gave us presents and money.

The most important visitor I had was Lord Snowdon. He came round the ward and he came to sit by me. He kissed me on the cheek and asked if I wanted to hear a story. He told me one. He was so nice. The nurses looked after me so wonderfully, especially a nurse called Karil. She was always laughing and trying to make me happy.

Weeks of pain and sorrow went by and I made progress. I had news of another visitor. This time it was a well-known television presenter named Alan Taylor. He had a puppet named Tinker, who we all knew from the telly, and he went around the ward, giving sweets out. I had a puppet of my own under my blanket and, when he came to me with Tinker, I brought out my puppet and tried to frighten Tinker. I talked to him and he gave me sweets. I had my photograph taken with him. This photo is on my mother's sideboard and it was in the *Merthyr Express* the following week.

I enjoyed the visitors, as many of my friends who had survived had gone home. (The devastation caused by the slide had been so complete that relatively few children were injured. Thirty children and five adults were treated in hospital; 11 were in for more than a week. One adult died in hospital.) By 5 November, when Dad lit the kids' fireworks outside in the hospital gardens, there weren't many of us left in the ward. Most had gone home.

Eventually, a young boy and I were the only ones left, and we would be there for many weeks to come. He had stomach injuries. The heartbreak of the disaster was still there but I was a bit too young to really understand. One afternoon, a man came to me and asked me what I would like to have. I could have anything I wanted. I chose a walkie-talkie doll. The man went down to town and brought back the most beautiful doll I had ever seen. This was a present from a school where all the children had given their pocket money. Weeks went by slowly. They seemed like years.

In fact, I was in hospital for three months. The iron stairs and lift to the ward were always very prominent in my dad's memory. The ward was always busy with reporters and families. Mam hated the crowds, feeling she could get no quality time with me.

*

Another of the nurses who cared for me on St Andrew's ward at St Tydfil's Hospital was Sheila Lewis, who worked on despite losing a daughter, Sharon, that day. The care I was given has always meant a great deal to me. I had been through a terrible trauma and was coming to terms with the loss of a brother and sister: spending months away from my parents was always going to be agony. I was desperately vulnerable and the caring hospital staff knew this. I wanted to talk to Sheila about those days and thank her for her sensitivity and skill.

When I met her she was still dealing with her own terrible and recent loss, the death of her husband, Gwyn.

'I often want to cry but I don't cry. I walk round with my eyes full when I look at Gwyn's photo. I just can't cry.'

We talked about the garden where Gwyn loved to work. Sheila had made a plan the night before we spoke of what she wanted the garden to look like. I said, 'Gwyn, there, will inspire you to do the garden, he will steer you through what he wants done.'

'Yes,' she said, and turning to his photograph on the mantelpiece: 'What would you like, Gwyn?'

I can see the comfort she gets from that photo. Then Sheila smiled. 'Gwyn, you have to talk to me and say it's OK!'

We both chuckled. Sheila said: 'It's good medicine to laugh.'

'What you are doing, it's history, and most importantly it's personal,' she said. 'I nursed you in hospital and I know how hard it was for you to lie in the bed, your father and mother coming to see you, and then you thinking that you should have died with the other two, your brother and sister.'

We both took a moment and then I said quietly: 'I remember that, Sheila. For many, many years the guilt of living was traumatic for me. I would have done anything to die and the others live.'

Sheila smiled. 'It's very hard to live with, and lots of children in the village had it. I don't think there were many that didn't feel that guilt. Even the mothers that had surviving children felt guilty because they knew that other children were gone and their children were alive, and they felt guilty about that, you know.'

I shared with Sheila my memories of living in Bryntaf, always out on the streets playing. 'We used to have dirty looks off the parents, who had lost their children, even as a child we sensed it.'

'Of course you did.'

'So we used to go and play away from the streets, not for the parents to see us and get upset.'

Sheila explained: 'People were out of balance, and it hit people in different ways, because people are different when things happen to them, they deal with it the best way they can. But they are out of balance emotionally. It affects their mind and health and also impacts on their body, and it is hard to live like that.'

Sheila said she remembered coming to my house shortly before the disaster. 'You were all small then and I remember your mother introducing me to you all, and I'll never forget it, your mother said: "This is my beloved son, Carl. My only son".'

Her memory made me almost choke with sadness. 'Oh, Sheila, bless him,' I said.

'I will never forget that moment you know. It wasn't that long before the disaster happened. Your mother, Gaynor, was so proud of you all.'

I asked about Sheila's own suffering. One of her daughters, Sharon, had been a victim of the disaster (she had been in my sister Marylyn's class), while her other daughter, Pat, had survived. Her son, Gwyn junior, had later died in a road accident. 'You have been bereaved yourself,' I said. 'I really don't know how people coped. I have children of my own. How did they ever cope? How, Sheila, did any of you cope and come through that trauma?'

'I think women have got an inner strength that they don't know they have got until they are faced with trauma and they have got to live through it,' she told me. 'You have to live through it and, if you have got children, and a husband, who had just lost the same as I, and he was absolutely devastated, so I had to be strong. I was strong and I buried it deep inside me and I did it because I pushed it to the back of myself. And when it came to the menopause, I had the most dreadful menopause because I had suppressed all my emotions to be able to give my children and my husband the support I needed. I didn't know what to expect, I hadn't gone through anything like that before. I was doing what I could to give them hope in the future.'

I told her that, while being strong for others, she was forgetting about herself.

'Yes, I had to live in the present,' she said. 'I used to come down from bed at 2am and cry to myself. I didn't want Gwyn and the others to know, so I cried when they slept and I would cry my eyes out downstairs, not for him to hear me or see me. It wasn't because I didn't grieve, I did grieve, but I kept on top of it. Because I had kept my emotions in, during the menopause I think I had a nervous breakdown. But I wouldn't have said that then. I couldn't take tablets. I said to the doctors I have too much to do, so I kept fighting for my family and the community.'

A fortnight after the disaster Sheila said she went back to work as a nurse on nights. 'I was nursing on St Andrew's ward at St Tydfil's. I was the staff nurse on night duty, two nights per week, Saturday and Monday, because I was better off working than at home, because I wasn't sleeping. I went back to work at the beginning of November, but I finished in April because Gwyn went downhill and wasn't sleeping at night and I thought, "I can't have him doing that. I can't look after myself, if he's suffering because I'm going to work at night and I am not there".'

Talking with Sheila took me back to my three months in hospital. I could remember the lift, the staircase, the décor... Sheila remembered the eight-year-old girl I was, grieving in a strange place, unable to go home. Sheila recognised that my wounds went deeper than a broken leg.

'You were suffering apart from that. I remember you in the bed, your eyes wide open because of the trauma. You never spoke much, you were very quiet. I never saw you crying in hospital. I thought to myself, "if only she would cry".'

I confessed that my mother didn't know how to deal with me when I came home. I wouldn't speak and became very withdrawn. No-one wanted to ask us how we were, so it was buried like it had never happened.

Sheila nodded. My experience chimed with hers, her memories of how Pat was affected.

'That's true. Pat didn't cry, but she was sleeping. Pat and Sharon always slept in the same bed, so I left her to still continue sleeping in the bed when Sharon died to see if she would cry. She didn't, so in the end I had to give that bed away to my brother because I just couldn't stand it. She wouldn't cry. Pat wasn't injured, she was the one that came home and told me what had happened. She was seven. Her teacher put her through the window and said, "Go straight home!" Gwyn was still there, he was across the yard in the infants' school. Pat came in, she didn't have a coat: "Mammy, sorry, I haven't fetched my coat, the school has fallen down, and I don't know where Sharon and Gwyn are." She was clinging to my arms tightly and wouldn't leave me. I wanted her to stay with a neighbour and she wouldn't.'

I told her about a friend who ran home in dirty clothes and told her mum that the school had fallen down, only to be told off for lying.

'Oh, I believed her,' Sheila said. 'I knew every child in that class, no-one came out alive.'

Sheila walked up to the school that day with the wife of Dr Jones. 'So we both went up the school, up the street, and a dreadful sight met us. I can see the steps going up to the yard. Gwyn, my son, was on top of the step; I will never forget the look on his face. He was like a sheet of paper, white, with one little tear trickling down his face. He said to me, "I don't know where Sharon is, and I don't know where Pat is, too". I told him that Pat was safe with me. He told me that he had tried to get their coats out of the cloakroom for them to put on. I said: "Don't worry about the coats, Gwyn".'

Sheila's husband, Gwyn, had come with men from the Taff Vale colliery to start the search. She felt she had to look for herself.

Sheila climbed through a window of the school and was approached by a teacher who asked if she was looking for somebody. 'I told her I was looking for my little girl,' Sheila remembered. The teacher told her: 'Get out from here! Get out from here!'

But Sheila would not leave. 'I was looking about and I remember looking on the wall where there were the paintings from thanksgiving the previous week. Paintings and drawings, I'll never forget that. I stayed there, I couldn't do anything. You couldn't get any further than the wall, the back wall, that wall was holding the rest of the structure of the school that was left, and all the children that died were behind that wall.'

That wall. My father always remembers that wall.

'I saw Dr Oliver and the men coming through the window and, when I saw them, I went out then. I knew they would work and do their job, and I spent my time going back and forth to the chemist for the doctor. My mother lived in Rhydyfelin and had heard on the wireless. She told my dad something terrible had happened. My father came over and I didn't see him. I was there till God knows how long. I felt helpless, I couldn't do anything, I was there to assist in bringing the children out.'

Some people who had lost their homes were re-housed on Perth-y-gleision park. Among them was a family who had lived in a small-holding on an embankment facing the tip. It had been in the path of the slide. My father told me that the parents who lived on the little farm had gone to town shopping with their son when the disaster struck. When they returned and saw the school they recognised scraps of wallpaper from their kitchen among the rubble in the schoolyard. They knew then that their small farm had been swept away, along with a grandmother and two of their children.

The chickens my father saw in the wreck of the school must have come from their home.

*

After the disaster, classrooms were erected in the yard of Merthyr Vale school. Sheila said there was no chance of children from this village going there, having to stand in that yard and look up at that tip. 'I went up to Mrs Jones and said: "We have to stop this, people are so upset," and, as I said, different people have different ways of showing trauma: some parents would say, "Get them out of the house because the houses were so sad", while others would say, "They're not going". We called on everyone who had lost children. I knocked on all the doors. People wanted to tell me everything that had happened. I tried to tell them not to tell me as I was feeling the same way about losing Sharon. I used to say to them, "I am so sorry, but please just say yes or no. Do you want the new school there?" They all said, "No". There was a group of us, one was Jean Gough, we held meetings in my front room, and we said we have to stop this, otherwise they will be calling on us to send our children to school and then we would be in trouble for not sending them to school. Our group of about ten people met with the education officer and we met him at that school. He said, "Come and see how things are at the school". I said, "I know how things are up there". They tried all ways to persuade us. Jean Gough said that the best thing for them to do was to take the new classrooms and put them down Cardiff jail because that's where they would have to put us, because we weren't going to send our children to school there.'

Sheila said that for the first two years these grieving families fought. 'Then people started to get tired and then people got ill and that affected the grief. A lot of grandparents died, my father, mother and Gwyn's father all died within the first two years after the disaster.'

Sheila said none of us could fill the space left behind in our lives. She looked at me: 'Gaynor, you lived through that. You got married, you had children, you have grandchildren, but you are stuck with that traumatic memory. We all are.'

Chapter Six

A Different Child

IT WAS GETTING near to Christmas 1966 but what a Christmas it would be. I had wonderful news. They were going to take the splint off my leg. I was frightened but pleased.

The nurse came with a trolley and started cutting it away. Everyone looked at me. I was afraid of not being able to walk again. She cut off the splint and my leg looked numb and white. I couldn't move it at all. I had two nurses carry me to the bathroom for my first bath in three months. It was so lovely to have the feeling of water around my leg, although my leg felt so lifeless. Days went by and I gained a little movement. The doctors helped me gently bend it. It was a small success but it really hurt. The feeling that it would snap again crossed my mind but my parents were so pleased to see me trying.

One afternoon, the ward cleaner, Katy, was cleaning under the next bed. I sat at the side of the bed before being daring and swinging my legs around to touch the floor, so I could stand. Katy was watching me as I tried my best, but I still couldn't get my leg to go forward. She told me to go back in bed; I felt helpless.

During the following week the ward was decorated for the festivities. Nurses came around in the night, singing carols. But we could not make it much of a Christmas atmosphere: we all thought of the ones that had died. My father and mother came to visit, and the Sister told my parents some wonderful news: that I would go home on Christmas Eve for one week, although I would have to go back to hospital to learn to walk again.

I was so happy and I couldn't wait for the morning to come. After a sleepless night of excitement, I was given a wash and then I waited anxiously. My mother came with my sisters and I hugged them all. I was desperate to be dressed and outside. My mam had brought me a violet velvet dress. It was beautiful and I wore it with a cardigan and black patent shoes. The thrill was in my parents' faces. My dad brought in a wheelchair and I was put in it. I could not wait to get started. I waved goodbye to the nurses and my friends, but I knew I would be back.

The Sister wheeled me out and I had to go down in the lift. I was giddy because I had not been out for a while. I got to the car and was carried into it. My wheelchair was put in the back. I sat useless and frightened of what I would see back home. I could not imagine what the streets of my home village looked like any more.

My parents acted as if they were strong but I knew different. We reached the bend to Pantglas Hill from where we could see Aberfan. It was just as if a mountain had fallen and flattened my village. It was such an ugly sight. I stared at this black mass, with nothing in the middle. I just kept silent. When we reached my street, all the neighbours were out to give me a welcome home and they came to meet me, gathering around the car. My father carried me out and I talked to my friends and I was taken into the house. It was a horrible feeling to get used to it. I crawled on the floor to go anywhere and it was a job to go to the toilet. My mam had to carry me everywhere.

Christmas Eve afternoon there was another surprise for me. There was a party down the Social Club for all the survivors. My parents took me down there in the afternoon and there were all my playmates – the ones that were left. I sat next to Elizabeth O'Brien, who had also badly broken her leg. She was still in plaster, and the both of us were unable to walk. I signed her plaster with my name. There was no talk of what had happened to us. We tried to enjoy the party as best we could.

Christmas Day was not like any other Christmas we had had before. In the years of innocence all of us used to stand on the landing and wait for my mother to call. We would all rush, pushing each other down the stairs, being so excited, and rush to our place in the living room where six different piles of toys were laid out with our names on.

Now there was no rushing or excitement. I had to be carried down with no brother or sister there, just Belinda and Michele. Sian was still a baby. There was no excitement any more, just immense sadness, but we had to go on and try our best to keep things as normal as could be.

Before returning to hospital, my dad took me up to the cemetery in a wheelchair, where I looked at my brother and sister's graves. The cemetery at Aberfan is on the steep valley slope. A cold wind can blow there on anything but the brightest summer's day. I had later written about the day in my teenage journal and noted: 'On the way up the hill, Mrs Dinnage came out and put a blanket over my legs. As my father pushed me up the hill nearer to the graves, his voice got choked and quiet.' I looked around to see the white stone crosses that marked the deaths of all those innocent children. Most of my friends were buried in a line, all close to each other. That's the way we all were as friends. It was so hard to take in.

I began to feel better as the days went by but I did not know what my father felt as our family seemed shattered. The dark frightened me. I had to sleep with my parents every night. I had started to wet the bed and this was very difficult for my parents to cope with as I would not sleep on my own. Many nights I would wet my mam and dad. God knows how they put up with all these traumas that came their way.

I owed everything to my grandpa as he saved my life and I would never forget what he had done for me. We had a bond which was stronger than anything.

I can remember when I was helping my mother peel potatoes for dinner, my gran forgot I could not walk and sort of pushed me. I could not get my balance and my leg never moved to save me. I fell straight on the floor, crying I would never walk again.

The sorrowful week went by and I had to go back into hospital. I hated the thought of going back but I had no choice. The nurses all greeted me, and I still had the same bed. It was a Tuesday afternoon. All the patients were sleeping. Katy was again cleaning and I asked her to stand over the other side of the bed, as I stood up and put one foot in front of the other and reached to her. I put my arms around her and I felt so happy. The Sister came and gave me a row that I had got out of bed but I showed her I could walk and she hugged me too. I had a limp and walked very slowly, but it was a beginning. I later wrote in my journal: 'When the doctors came around the next morning, they were amazed at my determination and, if I progressed, they would not keep me in hospital for long. I walked to my parents when the visiting time came and my mother wept. I knew how she felt.'

I had been in hospital for 12 weeks, 12 weeks of agony, sorrow and pain. Then it was time to leave for good. I was sorry to leave the nurses who had been so nice to me. They all wished me well and said goodbye. The nurse that I had got most friendly with was Karil. She was such a jolly person, always laughing. I would be sorry to leave them, but I felt so happy as I reached my home and again I was greeted by everyone. They all looked at me walking and they were pleased.

*

I was home but I was not the same child. My parents told me recently that they could see that I had changed. I was very withdrawn, very shy, very quiet; that was the worst part for them, the silence, not knowing how to handle me, what to do or say. I had been taken from them for three months, so the family was trying to come to terms without three of us really.

They never dared to ask me questions, as they were too afraid to upset me. 'Physical injuries can heal,' Mam said to me when I spoke to her recently. 'But mentally... You wouldn't

My siblings sitting on a window sill at our family home in Bryntaf before the disaster. L–R: Belinda, baby Michele, Marylyn, me, Carl. Sian was yet to be born.

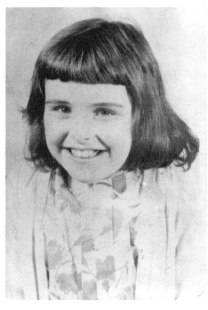

My sister Marylyn.

My brother Carl.

Carl at the beach. That is me peeking around the corner.

Marylyn holding baby Sian. This was taken not long before the disaster.

A family day out. Dad loved his cars. L–R: Marylyn, Carl on the bonnet, Belinda standing, with Michele (in red) and Mam behind and Dad in the driver's seat.

David Hughes and his dog Shep
in the mountains.

Enid Hughes

Me, aged 6.

This photograph shows the tramway which took coal waste to the tips at the top of
the hill in the background. Pantglas Senior and Junior Schools lie to its left.

Old Merthyr Tydfil

The houses of Pantglas. This photograph was taken two months before the disaster, in August 1966.

Trevor Emanuel

The view of the tips above Aberfan, from Bridge Street.

Trevor Emanuel

Pantglas School, Moy Road, Aberfan, in 1900.
Old Merthyr Tydfil

The teachers of Pantglas School. This photograph was probably taken a few months before the disaster. Back row, L–R: Howell Williams; Marjorie Rees; Margretta Bates; Mair Morgan; Doreen Oates, a supply teacher. Front row, L–R: Hettie Taylor; Ann Jennings, headmistress; Arthur Goldsworthy, who retired in the summer of 1966; Rene Williams.
Howell Williams

The last photograph to be taken of my class before the disaster. I am in the middle row, third from the left.

Gerald Kirwan

After the disaster, the school clock, stopped for good.

Media Wales

The front of Pantglas School. A classroom is buried under the rubble.
Old Merthyr Tydfil

Two children look on at the devastation caused to my classroom (centre) which took the full brunt of the slide. It's hard to believe that I survived this.
Media Wales

James Bullock (far right), 1st King's Royal Border Regiment, who, with many other soldiers, helped with the rescue operation in Aberfan for three weeks.
James Bullock

Bare hands and shovels search for life and clear the devastation.
By permission of IC Rapoport

Enid Hughes and her baby Tina in the street with shocked villagers. They are staring at the horror unfolding in front of their eyes.
Enid Hughes

Sheila Lewis holds a poem composed after her daughter was killed in a rear classroom of the junior school. Her son Gwyn, who was in a front class, survived.
By permission of IC Rapoport

This girl survived but lost her brother in the disaster and now clings to her father's side.

A man who has lost his wife, two children and home reflects on the future.
By permission of IC Rapoport

Where have all the children gone? Few play on the streets any more.
By permission of IC Rapoport

A mother mourns over her daughter's grave.

By permission of IC Rapoport

In the shadow of the tip: with my sister Michele (middle) as Dad comforts us while we look down on the graves.
Life Magazine

Mam tending to Carl and Marylyn's graves.

Me in hospital, with Alan Taylor visiting and entertaining me with Tinker and Taylor.
Media Wales

Other survivors, their families and nurses on our ward during Alan Taylor's visit. I am in bed on the left. My sister, Michele, is in the front in the check dress. It is just before Christmas 1966.
Media Wales

Me in a party hat with my sister Belinda on Christmas Eve 1966. I was still recovering and unable to walk at the time.

The survivors: a school trip to Llangrannog. I am in the middle row, third from the right.
Howell Williams

go to bed or be on your own.' She doesn't remember me crying much. I became very isolated from them.

I missed my friends dearly and most of all my brother and sister. Carl was the only brother we had, and now he was gone. Sian was just too young to know anything.

I had lots of money given to me and I went to lots of parties with my friends that were left. But I guess it would take a very long time for my parents to be happy again. It was very hard for a normal eight-year-old girl, full of happiness and used to doing all the usual everyday activities with her family, to deal with such a catastrophic change, which struck suddenly but would change everything forever. Each member of my family felt deeply disturbed, having feelings that were too difficult to deal with. I guess that more often than not each of us did not know how to cope with having to wake each morning and face another dismal day of misery and grief. How could anyone so unprepared deal with this burden? As a close family, every day was a struggle, each of us trying to give a little of the hope we had left to make each other's lives worthwhile.

When I thought of my brother, sister and all my friends, I tried to lean on my faith. They had all gone, I told myself, but I knew they were safe in God's hands, and they were being loved in Heaven. But such tragedy is bound to bring questions, even doubt. How was it that, knowing we had gone to church every Sunday, God could let this horrible nightmare happen to the innocent? How can you regain your prayer when you feel He is not listening, just when you need Him most?

But, as time went on, the only thing that kept me going was to believe again and ask His help to see me through. Wherever I was, whatever time of the day, I could talk to Him, bringing my emotions out of me, and I always felt better afterwards. For me, my faith in believing I will be united one day with those that I lost has got me through many of my darker days.

However, when I asked Dad about faith he became angry. 'You're different to me, Gaynor,' he said. 'Anyone with a sound mind that went down that school that day, they are not believers,

I can tell you – no way, no way! The thought of many strong men crying, howling, many collapsed in front of you. Bloody terrible sights that I will never forget. I have gone through too much.'

My parents remember a group of Mormons who brought toys and clothes for the kids around the houses. 'We threw them back, saying "We don't want charity",' Mam said. Nuns came around, then the Round Table. 'I don't want your money,' Mam told them.

We always believed in God before the Aberfan Disaster, but when asked whether she believed now, Mam jumped down my throat. 'Of course not. We never went to church after that.'

She remembered a conversation with a good friend, Nan Broad, whose son survived the disaster. 'Nan Broad, she said to me, "God gave his warning, Iris," and I said: "No way. It was man-made."'

'Nan Broad said: "I can see what you mean".'

Mam said to her: 'If God was so great, he could have parted the sea of slurry like Jesus and saved all the children.'

When I heard this from my mother I was very disturbed. I suppose Mam is right really. Why couldn't God have saved the children?

I paused in conversation with my parents and then asked them something which had been on my mind my whole adult life. Had the horror of what happened, the grief and the struggle afterwards, put a strain on their marriage? No, they both said. It made their relationship much, much stronger. It completely unified them.

As the breadwinner, Dad was always working in the quarry which he owned, making slab stones and selling them, picking up small coal from the gullies around the village. His chickens gave us eggs, and all this kept him busy and occupied his mind: he had to have something to focus on. Dad was also a fighter: busy fighting for the rights of Aberfan.

Mam couldn't work: she was too ill for many years. To my shock and horror I learned Mam was prescribed from the day of the disaster, six Valium each day, each tablet containing 10mg which went on for many years. Mam said she felt doped all the time and stricken with grief. The effects of all this reduced her to skin and bone – five stone – within five years of the disaster. She looked dreadful.

We all fell silent in our conversation. Then Mam explained that she had tried to stop taking the tablets but that she was now still on them, 50 years later. She had tried endlessly to lower the dose but she felt that she would never be able to be rehabilitated off them. 'It's too late,' she said.

My parents explained to me that they had four beautiful girls who kept them going. After the disaster, they found it hard to be parted from us. Gran Minett used to take us girls down to her house to give Mam a rest but, as soon as we had gone, Mam said she felt the overwhelming urge to have us back, not wanting us to go out of her sight.

Always feeling the immense pain of losing my brother and sister, I am shocked to hear my mother revealing a conversation which took place with a psychiatrist at St Tydfil's Hospital. She asked him: 'Can't I have another child? Another child will help me.' She was pleading with the doctor to agree with her.

The doctor told her: 'Look, Iris, if you have another child, it's not going to make up for the two you lost.'

Carl was her only son. When she thinks about him, she knows she could never replace him. Marylyn was an individual too and, as a mother, she had hopes and dreams for them. It is so unfair to think of their lives unfulfilled.

*

The bungalow where Mam and Dad live is named after Carl and Marylyn: 'Carlyn'. The foundations were only built at ground level when the disaster happened. After that day, Dad just didn't have the heart to finish it. He said he did not even want to pick up a brick for more than four years. We all, as children, helped Dad build it, even though we were so young. We carried bricks. It meant so much.

At one point my parents made a decision to move away, anywhere from Aberfan. But then Dad recalled saying to one vicar, 'What's the point in running if it stays with you?'

'He said: "Yes, you can go where you like, but the memory is always there."'

My parents must have been so desperate, confused with nowhere to run away to, to escape their despair and sadness. I am so glad they stayed and we fought this fight together as a family. That solidarity makes us what we are today – strong.

Chapter Seven

Back to School

THESE WERE STRANGE times and, with all the issues left by the disaster, I always feel that I missed out on my childhood. Most of my friends of my age had died and this left a wide age gap among the survivors. But we all struggled along each day with hope that time would heal our pains.

When it was time to go back to school, I was sent to Darrenlas Primary, which was quite near the railway track in Mountain Ash. The trains were heavy coal locomotives and they rumbled regularly past the school. I remember the first train that went by while I played in the yard. This had a traumatic effect on me. The heavy roar of the wheels on the track made a similar noise to that terrible day. I stood still with fear and held my hands up to my ears and cried. As if not enough had happened to me, the position of Darrenlas made it impossible for me to bear and I had to move school again because of the noise. This time I went to Troedyrhiw further up the valley towards Merthyr, and had to begin to make new friends again. Everywhere I went, pupils looked at me as if I were an alien.

For all of my school years, I was either in the doctors or in the hospital for one thing or another. My school work was immensely affected but teachers never pressured me. They always felt pity for me, more than anything. After my leg mended, it made my damaged leg one inch shorter than the other and I had a limp, which I hated. My mother had to take me up the hospital three times a week for treatment. My leg was put into wax which made it easier to manoeuvre. After that, I had to scrunch cotton wool tight between my toes and walk up and down the length of the hospital room. What a task my mother had to be put through. She must have been tired and still grieving over her loss but never once did I hear her moan. I had problem after problem to face. My shoe had to be especially made up so that it looked like an extra platform to level my limp. 'Why me, all the time?' I kept asking myself, but I was alive.

How did my parents cope with such a battle, the will to go on? What better future could they dream of to give them strength to survive the years that lay ahead? They had so many emotions to deal with and all the while bringing up a young family.

*

By the time I was ten, I had made new friends. They were older than me, 12 to 14 years, but they were company. I was good friends with Robert Holt, Paul Carpenter and Susan Maybanks. We went for walks and camped sometimes. My friend, Gillian, came from Birmingham. We slept up in the woods, me, Gillian and Michele, my sister. We slept in our old tent which had sides that didn't even touch the ground. Paul and Robert slept in the other tent and we were really frightened. Many of us tried to forget the disaster but it was impossible as the question kept on coming up, time and time again.

My parents kept me occupied as much as possible, with several trips to the cinema. We were, as my parents said, 'little buggers' before the disaster, just normal children who just loved life and play.

It was when I got to Afon Tâf secondary school in Troedyrhiw that I began in earnest to write a journal. I got hold of a spare school book and, during every lesson, I pretended to be hard at work, while in reality I was writing my journal.

Why did I do it? I suppose I felt I was suffering in silence. I was living with grief all around me. It was overwhelming. I was grateful to be alive but, in the midst of so many deaths, survivors can be kind of forgotten about really. I just could not stop pouring out my heart into the little book. I hated it when each lesson ended as I wanted to stay writing; there was something inside me driving me on. Wherever I went, I just opened the book and kept going.

I also wrote poetry when I was feeling isolated and had nobody to turn to. I thought that if I told people I was suffering, they would laugh; after all, I was only a child. In hindsight, I now know, that this writing down of my feelings has been my own way of healing. After all, it is what I am still doing here. Many who survived were not so lucky and did not find a way to deal with their trauma. They could not cope and died of a broken heart many years on. One solicitor who represented villagers wrote in the weeks after the disaster about one bereaved father who would at times go 'into a room on his own and sit down and just grieve, and he was then in agony trying to visualise what his wife and his children went through before they died... [This] is what is happening in dozens of homes in the village.'

The most appalling feature of the village's grief was that everyone suffered. This meant that there were too few left unaffected to give comfort and strength. The whole village was trapped in loss.

*

There was talk amongst the survivors that their parents had letters of invitation for each child that survived to see a psychiatrist. I was totally horrified to hear this from my friends. God, what would people think of us seeing a 'shrink'? I confronted my parents and told them bluntly there was no way I was going for these head tests and talks. A few weeks went by and, as far as I knew, I did not get a letter. Susan Maybanks did and I questioned her when she came back from her appointment. I did not know at the time that the appointments had been arranged in order to assess us so that we could claim damages; I just thought it was a very disturbing thing that us children were being put through.

Hadn't we been hurt and damaged enough? I was very adamant that I was not going. My parents tried their best to reason with me, and told me of the importance, but each time I got very upset at even the thought.

Two weeks went by and my parents said they were taking me on a shopping trip to Pontypridd. I put my best clothes on and we were off on our journey. Halfway there my mother told me we were going to see the psychiatrist. This was the only way that she could get me to go. I was horrified, crying and shouting, 'I am not going'. I felt that everyone around me thought I was mental.

I can still see the hill to the hospital. The car stopped. My parents tried to get me out of the car. I could not stop screaming. They finally managed to get me into reception. I was hanging on to a radiator for grim death, trembling, my mam struggling to try to get my hands free. My emotions exploded. What were people thinking?

A nurse came to try to calm me down. There were patients wandering around. One came over to me, mumbling something. God, I felt just like them. I felt humiliated. Cuthill was the doctor's name and I could not believe what a calm approach he had to me.

I was able to talk to him with the greatest of ease and tell him all my fears and feelings, which helped me immensely.

The head examinations were like some space tests with wires and metal clips gelled to my head. Lights were shone right in front of me. It was as if I was in a trance. No-one can imagine how I was feeling. As far as I could work out they were looking for brain damage or any effect on my state of mind. Looking back, my physical and mental state was so disturbed, it was no wonder my parents were concerned about getting me well again. These appointments were regular over a period of a few years. Dr Cuthill helped me in many ways. I owe him my gratitude. No-one should ever be ashamed of seeing a psychiatrist, child or adult.

During one visit to the psychiatrist, Mam recalled that I was very quiet. In the waiting room there was a woman who I thought was the Queen. Looking back now, I must have been such a naive child – most 12 year olds know what the Queen looks like.

*

Getting older, I was beginning to brighten up. I went to many dances and could have had many dates, but I was a shy girl. I was afraid of boys, but liked talking to them.

Things were starting to brighten up in Aberfan, but mourning still went on. My mother always said she couldn't go on if it wasn't for us four girls, as we gave her strength. Every anniversary of the disaster left a weeping sorrow and my friends cried and also my parents, as we could never forget. I missed my brother and sister very much but I hardly cried. I believe in God very strongly and I know He's loving them as to me they're still alive but in another world and this has put my mind at rest.

*

I was a quiet teenager and always dressed smartly and kept myself well and I had many compliments. I still had to go to the hospital but monthly now as they had to check on me. My visits to the psychiatrists at Church Village were traumatic. At 14, I was very aware of what the old hospital stood for and it was by then being talked about by the survivors as 'a mental hospital'. I didn't want to be seen there: this was a big stigma for a young teenager to carry and people looked at you differently. 'I'm not mental,' I said to myself.

When I talked with my parents in January 2015, I was shocked to learn from my mother that both my dad and mam made the decision to take me there. She had received a letter from the Coal Board offering the option for me to go. When she tells me I feel awkward. I giggle with embarrassment. My mam seems to imply that it was an obvious decision, that I was showing telltale signs of mental health issues and post-traumatic stress. I was wetting the bed, I did not want to go anywhere, to bed or back to school. She recalled taking me to the doctor in Troedyrhiw and a train passed over the railway bridge when I was passing underneath: I was frozen with fright and cried; once again, the noise reminded me of the disaster.

Our local GP, Dr Arthur Jones, told my dad no-one could tell him how long the effects of all this would last. I can understand why Dr Jones said that. My parents never knew what I had written in my diary until I was 14, when they took the little blue book to the National Coal Board psychiatrists, who would assess how much compensation I would get.

*

Susan Maybanks and I went dancing and we were very close. We got in a gang in our street and played football or rounders which was fun and we all used to enjoy ourselves. We both used to go to a new school of temporary classrooms that they had built in the village. But, when they built Ynysowen Junior School, I went there for a year.

After school we went to the Navigation hotel, where Peter the farmer left his tractor. We went up to the farm with him every day and had some fun. I like farming and wasn't afraid of the animals, except the bull which chased us many times. At weekends we would help Peter and his brother Donald do their chores around the farm. Donald had a son, David, and Susan and I were always around him. The three of us were great friends but Susan and I were both jealous of each other as to who was to have him as a boyfriend. I really got to like David and I wanted to go out with him, but I was shy.

One Sunday afternoon, Susan had not come up and Russell,

David's friend, was there. I asked him to ask David to go out with me and he told me to go and see him. Again, I was too shy.

Then Russell and David came to the gate where I was and that was where I had my first kiss from him. I blushed so much but I was really happy that Susan had lost!

From then on, I went out with him, although we were really just good friends. The more we spent time together, the stronger we courted. We had rows and finished sometimes but we always ended back together the following day. I really learned to love him but if we had a row I would sometimes cry. I was the sort of girl who would be sentimental. Sometimes I had a feeling that he was going off me but we stayed together. I fell in love with him but was not sure how he felt about me.

*

Sometimes when the night came and it was time to go to bed, the memories of my past would go through my mind and I would cry pitifully. I tried not to let my parents hear me, but the more thoughts got stronger in my head, the louder my cries and these were heard by my mother. She came into my bedroom many times and always comforted me. There was nothing else she could do for me as she could not imagine what thoughts were going through my mind.

Worst of all, I had this terrible feeling of guilt. Why couldn't I have died and my brother and sister have lived? This feeling stayed with me for many years, and many tears were shed on my pillow. In the morning, when I got up, I always felt embarrassed that I had got upset.

*

My father had had plans to build a bungalow before the disaster and I just couldn't wait to live in it. Carl and us girls had helped Dad carry bricks for the foundations. But after the disaster

Dad had lost heart, and work on it had slowed. However, we then all got together and made it happen. It was to be built on ground next to David's house. It was a big job to move bricks and carry things, but it wasn't long going up. I liked the idea of living next to Dai as I could get to know his mother more. After the bungalow was built, we all got anxious to move in. Everyone helped to move things. The bungalow remains the family home.

David became a key figure in my teenage journal: 'Dai came to call for me every night and sometimes I went over to his house. His mother was nice and I liked her. She was really good to me.' David and I had been courting now for a year and by this time I loved him dearly. I began to love him more each day, but in a way it was not good for me as I'd be so broken-hearted if we finished.

I went to his farm which was up a hill. It was an old shabby farm but that's why I really got attached to it and it was great to get away from the streets. It made me happy and I dreaded the day the farm would be sold to the Forestry Commission. Donald, David's father, was a strong, tall man. He loved farming. I did as much as I could on the farm, and I remember the sheep dipping, and the shearing. We all had great fun. I'd be covered in muck but I didn't care. It was the only thing that made my life happy. Donald said I should have been born a boy, as it was strange to see girls grabbing sheep, and that I had the muscles of an ox. I only wish people would know what they're missing in farm living and the great fresh air. We went to sheep trials and cattle sales. It made me feel a part of it all and I felt that some day I wanted to be a farmer's wife.

David was the world to me. But we had a few arguments and we finished. The word 'finished' broke my heart. I had made a terrible mistake. It was over. I'd gone out with another boy to try to get him to prove that he loved me, to drive him back out with me, but it didn't work. He told me he never wanted to go out with me ever again. If he really meant this, he never loved me like he said. In my journal I wrote: 'I cried until I made myself sick, until I couldn't bear it any longer.'

I wrote him a letter and gave it to him. I waited up at the farm, the place that brought back memories of where we were brought together. I started crying by myself and Sam, the horse, looked as if he understood. David was down the street reading my letter. Then he came up, threw the letter at me and said: 'I don't keep trash and everything in it wasn't true.' My eyes filled up and I told him that I loved him and he caught hold of my arms and said his love had died for me and I replied saying, 'My love will never die for you.'

I sobbed and my head ached. The tears ran down my face. He said: 'Gay'n, it's goodbye here.'

He kissed me so gently but I couldn't stand it. He was going. I wanted to run away that night, and I cried until I was ill. He had to practically carry me home as I wouldn't go.

I wrote in my journal: 'I went by myself to the cemetery, as every time I'm sad it seems to make me feel better, somehow. I believe in God and if I pray for something I seem to get it. I don't see my brother and sister as being dead, nor the others. I see them in another world with all white dresses on and God taking care of them. And when they're smiling, we are happy.' I stayed in the cemetery for one hour and by that time I felt well, but his name flowed through my head like running water. On and on.

I went home to bed. I prayed he'd come back and promised I'd never like another boy. I longed for that night to go so that another day could start. The next day, I went for a walk as the sun was glorious. Faulky, my friend, asked me if we were back together. I told him we weren't but I could see David walking up the hill. Faulky then talked to him and, after a moment, David shouted to me, 'I'll see you at seven o'clock.'

I didn't believe it after what he had said, but seven o'clock came and I waited up the field. He came and my heart lifted. He asked me to go for a walk and he pressed his lips hard against mine. He whispered, 'I love you', put his arms around me, sure to never let me go. And from that day on, we got even closer together.

*

I felt so free when I was up on the farm, especially on the mountain which was known locally as the Tumpy. The trees hang over each other and the green fern, oh, it's breathtaking just to look at it. When I was a young teenager I wrote: 'I'd love to have a log cabin and a small farm in the middle. I would very much like to marry David and have his children and in God's name I hope this dream might come true.'

I had plenty of guts in me and my mother always laughed at the strange things I did. One secret though I kept from them. David showed me how to use a 12-bore shotgun. I would sit on the farm gate and see how many blackbirds I could shoot. One afternoon, Donald and I found a ram with its horn growing into its eye. We held it still, and used a saw to start cutting it off. I was surprised to see blood spouting out of a small hole which runs through the middle of the horn. To stop the bleeding I had to scrape my hand across the top of the barn door and get a handful of cobwebs. This made me feel itchy and dirty. It was then stuffed in the end which stopped the bleeding.

Ear-marking is a job of guts, too, and I had to hold the lamb up against my chest while Donald snipped the top off the ear and made two splits down. I could hear the squelching of the ear. Some bleed but some do not. This method was cruel but all farmers had to do it. I have seen hog pigs being done, which not many people could stand, but I used to say that I had 'the guts of a man'. A sow had piglets and I watched, but it did not hurt me in any way as I felt the sow should be proud after all the pain she went through.

As a child I wrote: 'I hope when I'm 18 or 19, I will have the honour of giving birth to my children and I couldn't care how much pain I go through as no pain can hurt me as much as my past. That's one pain that will never be forgotten. The agonising torture of death and sorrow, which will dwell through this world for ever and torture minds of the lonely citizens of the little village of Aberfan, a name never to be forgotten.'

I would eventually have three lovely children, James, Ben and Cassandra, although not with David. At 19, I married Kevin, a postman, although we later divorced.

David and I had split up when I was 16. Sadly, he later suffered a medical problem which made communication difficult. His mother died of cancer. He remained living with his dad, Donald. The farm where I spent most of my days as a teenager has been sold now, but there are still the memories, of the hillside, helping Donald and walking with David, hand in hand.

Chapter Eight

The Engagement Ring

DAVID HUGHES'S SISTER is Enid. She was born in Perth-y-gleision Farm on 12 May 1945. She was 20 at the time of the disaster. She was due to turn 70 the week after we met to chat about David.

Enid had had polio when she was only 18 months old and had spent many weeks in hospital. It left her lame on one side. She'd had to have the leg amputated.

She lived up on the farm above the village. She loved the horses and farming itself. As well as David, she had another brother called Cliff, and an older sister, Jo. Her father was Donald, her mother also Enid.

As well as young Enid's own illness, the family suffered another early tragedy when Cliff drowned in a pool at the farm. He was just two.

Her brother, David, was 11 years younger than Enid. His birthday was on 28 August.

Although it was 'mixed with a lot of tragedy' Enid had loved living on the farm. 'Dad would always make me laugh. Mam was always strict, she had to be. Happy memories of picking spuds. All the kids in Bryntaf would come up the farm and help us with sheep dipping. They'd play cricket with a stone. Health and safety wasn't around! We would go to jail now if we did the same.'

We both had memories of sitting on a tractor's trailer on the way home from school. We wore no harness as we travelled through the streets, the tractor picking up swill from the schools to feed the pigs. We held on tight and no-one ever told us to get off. Not even my parents saw the danger.

Enid said: 'When I had my leg off, I couldn't go far, so I wanted to go out. I was on the tractor. I thought I would go to a friend's and the tractor took off with me on it. I came around the corner by the cemetery and I turned the tractor into a rock to try and stop it. The tractor came over the top of me. I wasn't injured but I was taken up the hospital. I never drove the tractor again.'

Enid told me about Pantglas in the years before the disaster. 'Lovely memories,' she said. 'I loved sport and the teachers. Muriel, my teacher, I was only talking to her a couple of weeks ago. She was the youngest at the school. One teacher at Pantglas had gone away to buy paintings for the school. One painting was called *The Blacksmith* and it fascinated me. On many occasions I would stand and just stare at it. If anyone wanted to find me in the school, they would say go to the painting. Every one of the hairs on the blacksmith's arms had been painted individually with different colours. Where that painting went I do not know. I looked everywhere for the painting after the disaster. I am not one for art, just that one painting. It was hanging in the senior hall at Pantglas.'

As Enid was older than me, I asked if she ever heard anyone voicing concern about the tips? 'I used to look at the field when I was going to school and think to myself that the field is getting smaller and smaller every year. I didn't see the danger of it, being so young. My headmaster's wife, Liz Evans, always said it was going to come down...'

*

She said David and her had a very good relationship, although they 'used to fight like hell'.

'When I look back I think Davey always had that illness. He always had headaches, his friends would tell me he was always on tablets. I think he thought his headaches were a natural part of his life. We took him to find out but nothing was found. They took six months to find out it was MELAS

105

Syndrome, which there is still not a lot known about now.' [It is a syndrome which affects the brain and nervous systems.]

Enid said: 'Donald idolised him, although they were always arguing. I can see them now.'

We both laughed.

Enid moved into the village when she married Terry. Her parents moved into a cottage as the farm had become too much for them.

Enid gave me a photograph taken in the street on the day of the disaster. It showed her holding her little baby, Tina. The look on Enid's face tells the complete story of what is happening around her.

Enid said: 'I don't remember having the picture taken. My memories of that day were very sad memories, Gaynor. Living in Coronation, the first I knew there was something wrong up there... I was feeding Tina in the front room window and a little boy opposite, Clive, had such a look of terror on his face that Tina dropped out of my arms. I jumped up and shouted for Terry, he was out the back kitchen. He hadn't gone to work that day. "Get up that school straight away, something awful has happened!" I ran in to Bob next door and he had Tina for me to go up there. I found out that Davey had been taken to hospital.'

David, who was in an older class that day, never spoke to me about what had happened, so I never really knew his story, and Enid said he never spoke to his family about it either. But, when he became ill, he developed a habit of going around to cut the hedges. 'He went up to the cemetery to do the hedges there as he thought it was his job,' Enid told me. 'If he saw a dead flower in any of the children's vases, he would take it out and put it in the bin.'

Her mother never told Enid and her sister Jo about what happened. 'We wanted to know because all these children that survived were going to suffer one day: I knew how I had suffered for all those years over my brother Cliff. It was such a sorrowful time with Cliff, and it was in later years that it affected me. I still didn't talk then, I have kept it in until now, but I wanted to talk about it with someone. Things like that do affect you later on in your life.'

I told Enid that it was only now that I was able to ask my parents things. 'I had to find the right time. I had to know before anything happens to them. I have to know.'

Enid nodded. 'I know David was trapped behind a radiator and there was something over his face. He had injuries to his arm and shoulder.'

Was his later illness anything to do with the disaster? I knew he could become quite aggressive at times.

'Davey was a very deep boy and he never talked about the disaster so we didn't know how it affected him mentally. It could have well been a factor. We thought, as a family, David was just sad after the disaster, we didn't think it was a factor of his illness, and the hospital assured us later on in his life it wasn't.'

My fondness for David is still something I feel very keenly. I asked Enid about her memories of us and she laughed and gave a sigh. 'Happy memories. I remember going to the Isle of Wight, and David and you would be sitting in the back of the van – you would be saying, "Stop! Stop! I want to go to the toilet!" Every half-hour you wanted to go for a pee!' We both laughed. 'It was nice to see my brother happy as well, because David never seemed to be a happy boy to me. He always seemed as if there was something wrong but you could not put a finger on it. He was quite touchy. But my memory was seeing you both so happy, well that was my impression when you were in front of us. I was quite sad when you broke up.'

I couldn't help but open up to Enid. 'I wanted to have his children, be married, up the farm. I loved him. He was my soulmate.'

I told her that David never showed me his feelings, never showed that he loved me or told me he loved me. 'As children this was so important, so I went out with someone else to make him jealous, to see if he cared!'

This made Enid laugh. 'My plan backfired,' I said. 'He finished with me over that. David told me his love for me had died. For days I sobbed and cried to the horse, Sam, on the farm!'

I told her about Russell and my note asking Davey to meet at the farm gate. 'David met me up at the farm, by the same gate. He forgave me and gave me another kiss.'

David and I had got engaged when we thought I had fallen pregnant. It was Christmas Day 1974, and I was just 16. 'I remember going over to David's house with my mother, Iris, to discuss this… I was petrified.'

Enid said Mum was very strict and very protective of David.

'Yes, I remember,' she said. 'Someone in the family told me, I can't remember who. I said: "Well, if they are having a baby, they are having a baby, if they haven't anywhere to live they can live with us – me and Terry – till they get a home."' Enid laughed. 'I was married when I got pregnant with Tina, and I was petrified!'

David and I went to Samuel's the jeweller in Merthyr to get an engagement ring. We didn't tell our parents. 'We didn't really celebrate but we had lots of presents bought for us. They were kept over at David's house and, to this day, I don't know what happened to the gifts.'

I told Enid for the first time that I went on holiday to Pembrokeshire and I met a boy. Enid remembered why we split up.

'David came over to see me in my front room and I told him. We were both upset.'

Enid told me that David was heartbroken. 'Feelings are a lot stronger when you are young, Gaynor. Davey cried and cried.'

I let myself wonder out loud what might have happened if I had stayed with David. 'You wouldn't have stayed with him, Gaynor,' Enid said. 'I loved Davey with every hair on my head but he was hard work. There was something wrong. I don't remember David ever having a girlfriend after you.'

I had given David my engagement ring back. As far as I was concerned that was the end of the story. I moved on. I got married at 19.

Enid said: 'I came across the ring when clearing some things out from Davey's house when he became unwell. David never mentioned to us that he kept the ring. I had quite a shock when I saw it. It was in a little blue box, heart box. I couldn't remember what your ring looked like and when I opened the box I said to Tina, "This is Gaynor's engagement ring!" How did I know? It was very strange. Davey would never have sold that ring. He was never going to have anyone after you, Gaynor, don't ask me how I know, but I know he loved you.'

Enid told her daughter Tina she could keep the ring if she wished and, if Davey died, she could pass it on to me if she wanted to. 'I said, "If you want to give it to Gaynor you can, it's Gaynor's ring. I won't do it while Davey is alive but keep it for him. You can't do it now because Davey is still here." When Davey did pass on, she decided you should have it. Straight away.'

David died on 17 April 2005. He was 48. Enid gave me the engagement ring at the funeral. I broke down. It meant so much to me. If I was down to my last penny I would never sell that ring.

'I can see you now when that ring was given to you at the funeral, the look on your face. You started to cry and the week before the funeral you had said to me, "I don't cry no more". And I thought, "Oh Gaynor, if you only knew you're going to cry when you get your engagement ring returned back to you". This is such a powerful story, two people who are brought together in tragedy and Davey's ill-health. How he had kept the ring for such a long time.

'When Tina came up to give me the ring when Enid, my mother, had passed away and Davey, I thought it was a lovely touch as Tina too had loved you. You were like a big sister to her as she didn't have any sisters or brothers.'

I told Enid that I was unsure what to do with the ring when I died but I was coming around to thinking that I would like to be buried with it. She asked me if it comforted me to know Davey was buried near to the children. Enid said that the day Davey died she was told that he couldn't be buried with Donald and Mam. She was devastated. Her second husband Gwyn made sure the grave was able to support three burials.

Something came into my mind. One morning, my dad had said out of the blue that I didn't have a funeral plot. He said to me: 'You can either be buried with me and Carl or Mam and Marylyn.' I chose to be buried with Dad and Carl. My dad then felt at ease knowing I had made the decision. Dad wouldn't worry then.

I told Enid that it gave me a lot of comfort to know I am being buried with Carl and Dad. Mam is going to be buried with Marylyn. 'I will be near to the children, too, as I believe we will be re-united.'

Enid told me: 'I believe in God. I don't think I could have coped with all the things that have happened if I didn't believe. I believe in the Bible and you don't have to go to church to believe. To try and practise what I preach, I don't always succeed. I used to talk to God when Cliff died – there was no-one else I could talk to. I was only six years of age when he died. My memory of other things is vague but often I think it's strange: most children at six lose their memory or wipe out things they don't want to remember when they get older, but I can clearly remember Cliff.'

I said: 'I can relate to that, Enid. When we talk about the disaster now, I talk about it as if it was another child, not me. I have asked this question to psychiatrists and they say it is due to trauma and you displace the memory. A coping mechanism.'

Enid told me: 'What you went through, Gaynor, was not only your brother and sister and you, it was your friends, too, all of them. No-one should have to live through their lives the way you have. It's not natural, Gaynor. That was man-made, we can't blame our grandfather and father for that. They didn't know they were tipping on a spring.'

*

Enid and Terry divorced, and she had later remarried an old friend, Gwyn. Gwyn had only recently died when Enid and I met to talk. 'How things happen and change your life,' she said with a rueful smile. 'Gwyn, I loved him when I was a kid. He was always up the farm. Everyone thought that Terry and I were always thought to be together, but it wasn't to be. The week Gwyn died, I was so low, I just walked and ended up back up at the farm where Gwyn and I shared so many happy memories.'

We smiled at the memory of the farm. 'A lot of people came to me and said they had a marvellous life at the farm,' she said. 'When my grandfather was alive it was a booming farm, all the fields were planted. A proper working farm. All the kids used to be up there from Bryntaf, including Gwyn.' She smiled with fondness at the memory and shouted to Gwyn's spirit: 'I love you, Gwyn. Always.'

*

I wondered what the 50th anniversary meant to Enid?

'It's a time to reflect. A lot of people are going to go back to that day. Where have the 50 years gone? I lost an auntie and her little girl. She was my mother's sister. She was only 26 years old and the little girl, Cath, was three. I often wonder what their lives would have been...' Enid's words are choked off. 'The family moved to the houses next to Pantglas. I had a large family but I only have one auntie left. I always think about them, we were such a big family.'

I was thinking of how religion had helped Enid cope. 'My parents are now atheists since Aberfan,' I told her. 'I find that hard.'

'My mother went through that stage but she came out of that, that I know,' Enid said. 'Aberfan needs to just stay all close together and give each other a hug. A lot of people said the disaster split the village, I couldn't see it. I was 21 then, I was young, I don't remember people talking about each other. I thought a lot of the bereaved pulled together quite quickly. I can't remember any animosity at all. I have heard about it since, but being young, I didn't listen to any gossip and don't think young people do. It's an older person's habit, that is.'

Chapter Nine

The Girl without a Home

I ALWAYS KNEW her as Francine Jones. We were eight when disaster struck. We were in different classes but lived a few streets apart. She was in Moy Road and I was in Bryntaf. We didn't know each other well, as we had our own sets of friends. But disaster was to bring us together and she has always been at the forefront of my mind. For years I had a strong nagging urge to find her. My memories of her had plagued me, in a way, throughout my childhood and adulthood. The little I remembered was that she was seriously injured like me. We both spent three months in hospital, Francine in East Glamorgan and me in St Tydfil's; we both took many more weeks to rehabilitate at home.

When it was time to return to class, our school had gone. As had many of our friends.

We went to the pre-fabricated temporary classrooms built in the park at Aberfan. Howell Williams and Hettie Taylor, surviving teachers, had the courage to continue to teach us. It was there that Francine and I shared a special friendship and we became inseparable. We shared a terrible bond of pain.

I remember both of us were unstable on our feet due to the trauma to our legs. Francine still stood with her leg in irons. We went everywhere together within the classroom. We were pitied. We never did any class work. I just remember it was always one of us who was asked to help wash the dishes in the kitchen.

Throughout our friendship I never asked her what had happened to her. We were just children. I dared not ask what happened to her legs for fear of upsetting her.

Then one day Francine had gone from school. Just vanished.

I didn't know to where, with whom, or whether she was ever coming back. I remember the feeling of despair and emptiness at the loss.

For years I had pondered where she went but with life's struggles I kept on putting off the search.

It was only when I started this journey that I discovered she was just under my nose: when I told my sister I was trying to find Francine I found that her contact details were held within a local bereavement diary.

As children we just did not speak about what happened, we blocked it out... I was about to find out what happened to Francine and it was to be a shock.

*

I headed to a street in Bridgend, looking for her house number. I asked a man, 'Excuse me, what number is this please?'

'Gaynor?' he said.

'Yes.' I was there. We laughed.

Then a tall, slim, dark-haired lady walked down the path. I immediately recognised her to be Francine. I felt deeply emotional already. Before she even reached me, I felt this bond.

'Hiya,' she shouted. Both our faces lit up with glee, amazement, surprise.

We hugged. 'You haven't changed!'

Francine smiled. 'You haven't either! You are looking well. Give us a kiss anyway.'

The man I had spoken to stood by the open garage door with a younger man. I assumed it was Francine's husband and son. They watched as we hugged again.

Francine's son smiled: 'Well, I hope you have changed: it's 50 years!'

We all laughed.

We moved inside her house. It was lovely.

'Are you still in Aberfan?'

'I will tell you about it now,' I said. 'Are you married?'

Yes, it was her husband, Roger, outside. He was painting the side of the house.

'I have been on my own too long,' I said.

'Oh, have you?'

'Yes, 28 years. Yes, a long time.'

'Never.'

'Have you always lived here?'

'We have been here 38 years now, soon to be. Yes, my eldest is 39 now.'

We were still smiling, giggling like the schoolgirls we were when we last met. It was as if we could not believe we were sitting there together.

'We go up to Aberfan regularly,' Francine said.

'Oh, see, it's crazy, isn't it? For such a long, long time as far as I can remember I had this vision of me and you in the classroom...'

'Yeah.'

'And, of course, I remember you had your leg in irons. There were bars each side of your leg.'

'Yes, that is right, yes.'

'And I had not long come out of hospital too, and we were like pitied on, weren't we?'

'Yes, I think we were at that time.'

'So I was always wondering what hope there was of me finding you. You were a Jones then and God knows how many of them there are and I didn't think of speaking to anybody, not even my sister about it. One day I happened to say I was looking for you and she said she had your number in the house. I thought, "Oh my god!"'

Francine laughed, and then gave a sigh.

'I had Mam and Dad's grave done and my grandfather's, of course. I went to see your father at the bungalow.' Dad was chairman of the memorial committee then.

As a child I had never known the full extent of Francine's loss.

I said: 'Do you know, I can honestly tell you I was shocked when I found out you lost your parents. And I am shocked now to know you lost your grandfather too.'

There was a pause before Francine spoke again.

'We lived next door but one to the school in Moy Road,' she said. 'We were 79 Moy Road.'

'I didn't know that's what happened to you. Because I was a kid, you don't understand. It was only because Belinda told me that I knew about your mam and dad. But I didn't know your grandfather had gone too. I am shocked.'

'Yes, my mam and dad and my grandfather, all three of them died that day at home.'

*

Francine, of course, had no idea of the horror which had struck her home that day. She was at school, buried under the sludge.

'I couldn't see any of my class. I was buried, thank God.' Both her legs were injured, her foot was trapped. 'Look at my foot now, it is still odd looking,' she said with a smile.

Just prior to the slide hitting the school, Francine had left or been sent on an errand out of the classroom. She could not remember why, but it had saved her life.

'I can remember coming back in, opening the door of the classroom and then BLACK. That is all I can remember. The door had come down on top of me, see? My leg, my foot, was all crushed. All the tissues and sinews, whatever, tendons, had all gone, so they had to cut me then from there to there.' She indicated from her ankle to above her knee. 'The operation was... oh God alive! And I have been back in since because I had all the coal dust embedded in my head and forehead. I had scars removed at 13 years old from my head. They couldn't take it all out, there is still a bit there. It was too near the eye, but it has faded over the years. I have some on my nose too, but as you get older, well, I don't take too much notice now.'

Francine paused, then went on: 'Do you know it's strange that four or five years ago I was having terrible pain here in my side and they couldn't make out what it was. They took me in, did a camera, X-rays, checked my kidneys, everything. I am now having my shoes built up in hospital as we speak. I have to higher this leg... I have been walking all my life without even knowing that I must have been walking on my side, lop-sided. So now I have to have special shoes made. This pain I was having this side of my body was compensating all my weight. It's from the disaster. And now they have found that out.' Only now, 50 years on.

Francine's medical condition was still affected by the disaster. I suggested to her that she was still in trauma really.

'People don't see the inside, do they?' she said. 'Oh, my life has been OK, thank God. My life has been pretty straightforward. I met Roger at 15, married at 17 and I had Lee at 18, and we've been together ever since. Forty years this year. Pretty straightforward. I met the right one, thank God.'

But there must have been difficult times when she was a teenager?

'I am not saying my teenage years weren't turbulent,' she said. 'We all go through that. When I look back, we couldn't have been sane. We never had counselling. It was, "Get on with it!", and I was eight at the time. When you were talking 14, 15, 16, that is a very hard time in life anyway but when you had gone through that, it has to be turbulent, you know.'

There had only been a handful of schoolchildren like us, the badly injured survivors; most of those who had been buried by the slide had been killed.

'We were very few siblings that were seriously injured and had lost close family,' I said.

'We were, weren't we?'

'I remember us together in that little kitchen in the park, wasn't it?'

'Yes, that's right,' said Francine, seeing us too.

'And then, all of a sudden, I remember your grandmother taking you away from me.'

'Yes.'

'And I never saw you again.'

'Yes. I was brought down here to live in 1968, with my brother and my sister, Eryl. My family was from Thomas Street; my father had his shop, the old grocery shop in Perth-y-gleision. Remember George James's shop? Further up from George James's there was a little corner shop on the corner of Cross Street. That was my father's shop.'

'I never knew that.'

'Yes, my father's shop. My grandmother was still in Perth-y-gleision but in 1971 I lost her too. She died in her sleep.'

'My God, all these years,' I said. 'All this time and your number is with my sister. She said you are always up the cemetery. But I called you Jones, I didn't know you as Bale.'

There was more to it than that, I supposed afterwards. We had a terrible tragedy in common. We had needed to move on before we could confront it. Confronting it together was very emotional. We had to look to the future now. I am glad Francine had the strength to break free.

In fact, when Francine told me her account she did it with extreme courage. She had never spoken about it before. As we talked we realised just how much our childhoods had shaped our lives. There was not only a feeling of nostalgia but also one of tremendous adrenaline; we were filled with energy as images of people and places whizzed through our minds.

But I was finding that making this journey was like an emotional archaeology. I could not just dig deep into mine or Francine's memories at first contact. We arranged to meet again, at a hotel in Dowlais, near Merthyr.

*

The second time we met Francine told me more about her family. She had three sons: Lee, who was 39, Damien, 37, and

Carl who was 35. She said she never finished her education because she had her family so young. She was working in a cleaning job now.

I showed Francine's some photographs of my children and family. We turned a page and there was a photo of my dad and Michele looking down at the graves.

'Oh, bless him,' Francine said. She remembered meeting Dad in Aberfan at a meeting to discuss her parents' graves. She had a fond memory of him being helpful and sincere.

Then through the door came Francine's sister, Eryl. We greeted each other and Eryl said she remembered meeting my father too.

Eryl said that their brother, Derek, was one of the oldest children to be caught up in the disaster. He was six years older than her.

'We lost him few years ago to Alzheimer's,' Francine explained. 'He was the one who brought me up.'

Eryl and Francine were living at home at the time of the tragedy; the two brothers – Derek and David – were married. Eryl was at work when the slide happened.

'I left home to go to work at 8.30am,' she said.

'Oh my God,' I said.

Eryl was working in Merthyr at the time. 'Fifty years it is, but it is like it happened yesterday. I remember everything. I heard all these fire engines and sirens going down. I didn't know what had happened. I was on break, this girl came running up the side of the lane and said the school had collapsed, that's all she kept saying. Until I got to Merthyr Vale and I could see what had happened... the whole lot had gone. My home, the school, everything, terrible.'

There was silence for a moment and then I explained about Francine and I becoming friends later and then her disappearing from school. I said I had not known until recently that their parents had both been killed.

'We all were shocked, Gaynor, because we didn't speak about what happened in the disaster even in the prefab classes in the park,' said Francine. 'No-one spoke about Moy Road either.'

Immediately after the disaster Eryl and Francine went to live with their brother Derek and his family in Aberfan Road. In 1968 they all moved to Bridgend.

'How did you cope then, after, as a family?'

Francine explained: 'My brother Derek and his wife Gaynor brought me up in Bridgend. Eryl lived with us for a short while, then she got married. I stayed there till I married Roger. Derek was 12 years older than me then, he had three children of his own.'

'How long were you in hospital?'

'About until January,' she replied. 'Off and on, then I had to go back in because of my leg again. I went back in then to have the scars and muck taken out of my head. The scar was the last bit taken out of my head.'

Eryl was 20 in 1966. She had been suddenly left without parents and had a young sister to look after. She said: 'I have taken a long time to get over it. I was the adult then.'

'You must have been a very strong family like us.'

'You just had to cope, didn't you? Get on with it. There was no help then. No help until this day.'

As someone who stayed in Aberfan I am intrigued about the fact that Eryl and Francine left.

I asked: 'How did you feel going away from Aberfan to Bridgend? Did you think then it was a good idea and what do you think now?'

'At the time I was only eight or nine,' said Francine. 'I thought, "Oh God we've got to go to Bridgend and live", but now looking back it was the best idea.'

'You were taken away from all your friends, your community,' I said, 'but looking at your life in comparison to mine, Francine, I feel that you have moved on much more than me because I stayed in the village.'

'You have got to put up with it every day,' she said.

That was true. Over the years there have been the injustices, the controversies, living with a father fighting for justice... It never went away.

'It's still going on,' said Eryl. 'The 50th anniversary next year.'

I nodded. 'I have never moved. Going down to Pembrokeshire yesterday I just hate coming back. I feel peaceful there. It is a depressing feeling. I am sorry now I never moved years ago.'

'I think that was the best thing for us then, moving out of there was the best thing for us at the time,' said Francine. Eryl added: 'We just wanted to get as far away from Aberfan...'

I said to Eryl: 'It is crazy. You were just going to work. Did you feel guilty too?'

'I felt I should have been in the house with them. A terrible thing to say really but that's how I felt.'

I explained about my own survivor's guilt, about wanting to die so that my brother and sister could be alive. 'I wanted to take their place.'

'Even I for a long time, for a long time, I used to think "Why me?",' said Francine. 'I think I was the only one who came out of my classroom alive. Why was I still alive when everybody else had died? I still think that: Why?'

I told Francine I found myself overcompensating for those who had lost children. I wanted to give back more and more emotionally. I made myself ill from doing too much.

'You need to step back,' said Eryl. 'You've got to. You still have a life.'

I admitted I felt my life had been 'all over the shop, no stability in relationships, in work'.

'It is most probably staying in Aberfan, not moving on,' said Eryl.

'Did you ever speak about what happened, as sisters?'

They both said not really. Then hesitated. 'Did we?' asked Eryl.

Francine said: 'I can remember being in hospital. We talked about that. When you came into East Glam I wouldn't let you go home. I can remember waking up and she wasn't there. She would probably go home in the night or whatever, and then I would cry and she would have to come back. That went on and on. We have spoken about that.'

Eryl, being that bit older, had been a good companion to Francine while she had been in hospital. It was something I had missed out on myself.

'I stayed all the time with Francine day and night,' she said.

'I couldn't stay on my own,' said Francine.

'I think you were so confused, that's why. It was hard. Every time I tried to sneak out when she slept she would wake and I had to go back to her side. In the end they gave us sleeping tablets, knocked us out!' They both laughed.

I asked Francine: 'Were you afraid to go to school?'

'I came from Aberfan and I went straight into school and I was OK,' said Francine. 'But what I would have been like if I had stayed living in Aberfan I don't know. I was fine in Bridgend, in Brynteg school.'

'Who told you your parents and grandfather had gone?'

Francine looked at Eryl. 'You told me.'

Eryl said: 'The doctor kept on saying to me to tell Francine and I kept on putting it off and off. Derek said to me, "You are the best one to tell her" because I was there with her at the hospital all the time. The doctor came in to me one day and said, "If you don't tell her I am going to tell her", so I had to tell her.'

Francine nodded. 'I know. I can remember asking for Mam and Dad. I remember they were going to visit David my brother in hospital. In my head, you see, they were going to see my brother, but they wouldn't come and see me. I remember saying about David not coming to see me all the time and I remember asking about Mam and Dad all the time. I can only vaguely remember being told about it... I tried to wipe that bit out, but obviously I had been affected.'

I said: 'I was told on the actual night that my brother and

sister had gone to heaven, and that was it. Then we never spoke about it until eight months ago.'

'Strange how no-one spoke about it,' said Francine. 'You just shut that bit off. Gaynor, you got it out in your diary.'

Francine said her parents' house in Aberfan was completely demolished. 'Everything had gone,' said Eryl. 'Derek found a few things and that's all we salvaged. Everything else went.'

'Did you have much help after to re-house?'

'No! Only what I was standing in, same as Francine. We had to start from scratch.'

'Oh, my God alive,' I said. 'Do you ever think how you came through these things?'

They both nodded.

'It's only when you have your own children you realise how awful this was. Were you protective of your kids?'

'Yes, I think I have been,' replied Francine. 'I think I have been over the top, not so bad now perhaps, but yeah in the past. Because Mam and Dad had gone, I looked at mine growing up and thought that my parents were not there at their ages, so what I didn't have I tried to give extra. My brother was marvellous bringing me up, but it wasn't my mother and father.'

She explained: 'There was only 17 months between Charmaine, who was my brother's daughter (she was in the front of the school that day and got out) and me. Then there was Yvette, who was in the infants' school at the top of the yard which wasn't touched. They lived in Aberfan Road. They ran home and I was brought up then with them, so I think I was there with them at the age of eight and we were like sisters, but I always felt, "but it's not my mam and dad". Although we were brought up together, that is not where I should have been.'

I said that although they had moved away it was good that they were strong enough to come back and visit.

'It is pleasure to see the graves all done,' said Eryl. 'They worked hard there. We had the gold lettering done on my parents' grave. Mam and Dad's graves are with the children.

My grandfather, who died in Aberfan, is down the bottom of the cemetery with our grandmother. She died in 1971. She was heartbroken. After the day it happened she didn't wear anything but black. She continually mourned. She would come down to stay for a couple of days and then say, "I want to go back, your father is waiting for me". It broke her heart.'

Francine said: 'I still go back and, I have to be honest, I enjoy going back to Aberfan. I love it there. Because in my mind they were my childhood years, you know. But it's strange, Gaynor, because, I said to Roger, I can remember being in the school, I can remember being buried and I can remember the door and me screaming for help; yet I can't remember much about going to school in Aberfan *after* the disaster.'

Francine looked down at one of the class photographs I had with me.

'Oh, gosh. You shut off, don't you? I can't remember most of the children in the photo. I can remember the night before the disaster, you know, but I can't remember leaving the house that morning and saying ta-ra to my mother and father. I can't remember that. I remember being buried, I can remember coming out of the school and being in hospital. I can remember the odd day in the Aberfan school afterwards but not much. I think that was because at that time I knew my mother and father had gone so you block out the bad bits... You block out the bad bits.'

Meeting up with Francine again had taught me so many lessons. To never take for granted the time you've been given with people who mean so much to you, as that time is an invaluable gift. When you rise in the morning, think of what a precious privilege it is to be alive, to breathe, to think, to enjoy, to love; make that day count. Because, as I feel more and more, you never know when this day might just be your last.

Writing Francine's story left me in tears. But seeing her again showed that our friendship had not ended, just paused. We were nine then, middle-aged women now; but the same people. Renewing that friendship was a gift to be treasured for the rest of our lives.

Before we parted at that second meeting, I looked at her and Eryl and said: 'If you had stayed at home that day you both would have been dead. The whole family gone.'

'The strange thing is,' said Francine, 'the night before we had all gone up to Bristol and we had come back very late. It was a half-day in school and Mam was on about whether to send me or not. She sent me. The family next door, they all died too with their baby. All the houses went. About ten houses in that row.'

Chapter Ten

The Bedrooms of Aberfan

MANY THINGS KEEP coming back to me. One is my dad pushing my wheelchair up the hill to the cemetery after I got out of hospital. His voice is weak; I am unsure what to expect. A woman comes over with a blanket and tucks it around me. That woman was called Evelyn Dinnage.

I have always just known her as Mrs Dinnage, a lady with blonde hair who as a child I considered to be somewhat posh. She lived in the quaint little cottage on the corner of the cemetery hill. I remember that when it rained heavily the water from the mountain would flood her home, so there were often sandbags piled high by her door. She was regularly in her garden, potting plants, while her husband Bill mowed the lawn. I remember the sound of their mower drifting across the side of the valley. The garden was always in splendid colour, with sweet peas and the biggest sunflowers I'd ever seen.

I did not know her very well. We probably only waved or uttered a few words but I would always remember that tiny moment of immense kindness.

Mrs Dinnage suffered some ill-health in recent years but her remarkable zest for life was inspiring to many.

As a young lady, she had worked in the wartime ammunition factory at Bridgend. She then went to London and worked in domestic service. She met the man who would become her husband, Bill, in Sainsbury's. He was working as a butcher. They married and had a son, Thomas. They moved to Winchester where Bill's parents lived and then to London, and finally to Perth-y-gleision in Aberfan and the little cottage near the cemetery gates. Bill worked as a medical representative and Evelyn stayed at home, helping Bill with his secretarial duties.

Evelyn is quoted in one of the *Life* magazine reports of the disaster. '[That] morning she could not see the town below at all, so thick was the fog. Nor could she see across the side of the hill to where the black tip stood. The power lines that passed her front door stretched in that direction, but she could only see a few poles away before the wires disappeared into the mist. Some time after nine o'clock she was startled to see the lines and the poles outside her front windows jump and shake vigorously, as if a giant hand were gripping the other end of the wires, whipping them about. "William, come look," she called to her husband. "The wires are shaking. What do you suppose is causing that?" They were still puzzling over it minutes later when the faint wail of sirens from the valley below reached them.'

Time and time again, I kept saying to myself that I should call over to speak to Mrs Dinnage. She seemed very important to my story. But there always seemed to be something in the way and months, years, went by. Discovering that she was now 94 and had spent some time in hospital, I knew I had to make the effort to see her.

I met Mrs Dinnage's niece, Sonia, walking her dog on the embankment by the cemetery just before Christmas 2014. I asked her about her aunt and found that Mrs Dinnage was now in a wheelchair and confined to her home. Sonia and her carers were looking after her.

I explained about the journey I was taking to meet people I'd come into contact with because of the disaster and that I wanted to talk to Mrs Dinnage to see if she remembered the time she comforted me. Sonia was quite emotional about the request and said both she and Mrs Dinnage would love to chat to me. Sonia had her own memories of that day.

On New Year's Day 2015, I met with Sonia and Mrs Dinnage in her little cottage. As I walked in she was sitting in an armchair looking really well. Her hair was immaculate and she wore a sparkly pink top. Her skin didn't show any signs of her age. To me, she looked just the same as she did all those years ago.

'My God,' I said. 'You look so well, so young.'

I gave her a big hug and a kiss. We laughed and joked about age, and how the years were catching up on us all.

'I hated getting old,' Mrs Dinnage said. 'You seem to be young, then all of a sudden where have all those years gone? I go to bed by 8pm now. When I sit by my back window washing dishes, I sometimes see your dad feeding the birds and I wave.'

'Yes,' I said. 'He often sees you.'

Mrs Dinnage told me that her memory was not what it was, but that she would try to help me. 'How old were you when the disaster happened?' she asks.

I told her I was eight. 'Brave girl,' she said, going on to insist that I call her by her first name.

'Evelyn,' I said, 'do you remember the time I was in the wheelchair, I had just come home from hospital three months after the disaster?'

She said that she did. 'I remember you coming up the hill with your dad. I was standing at the door. I remember the wheelchair with nothing over you. You stopped and we just started chatting.' She gave a sort of sigh, seeing the scene in her mind's eye. 'I thought to myself that your legs are getting cold. I said: "Wait a minute, I will get you something." I came in the house and got my blanket for you. It was a very cold morning.'

I asked her how old she was in 1966. 'I was 43 then,' she said. 'A spring chicken!'

Mrs Dinnage added: 'I remember when you started to try and walk again, coming down the path slowly, holding on to the walls. I used to see you.'

Sonia said she remembered going down the village after the disaster and standing outside Bethania Chapel. 'The minister came up to us – I was only 19 at the time – and he said, "Last week when I was here, I prayed that every seat in this chapel would be full following Thanksgiving. Now they are full with bodies of the deceased." I will never forget that statement from the minister.' She shook her head sadly. 'I used to walk up the cemetery hill many times with Christine and Avis – there was a 10-year gap but we were good friends.'

'Oh, Avis, she was my friend in school,' I said. 'She was in my class, blonde, pretty, always in beautiful clothes with a little ribbon in her hair.'

Both Christine and Avis perished that day.

Sonia said: 'After the disaster happened, I remember walking up the hill and thinking, right they are not keeping them from me. I wasn't a kid.' Sonia wanted to see Christine. 'I was determined. I insisted, they are not keeping Christine from me and I am going in there to Bethania. Where can she be? You can't just take a child like that. I had only spoken to Christine that morning and said, "Do you want a lift down to the school?" I was going to work in my car. I had recently passed my test. I stopped the car by the railings, it was foggy and I said to Christine, "Do you want a lift down to the school?" Christine said, "We are only going to school half-day because we've got Thanksgiving and I am meeting my friends." I said: "OK, lovely." It was just a normal day.'

Sonia hesitated in her story, and then she said: 'You can't help it. You just go a bit loopy, don't you, when something as tragic as this happens. I am a believer that when your number is up, your number is up, no matter what age you are.'

All three of us were mesmerized by our memories at that moment. I said: 'I always say that if I had not woken up, I would have not suffered and known nothing and, in a strange way, died peacefully without any pain or suffering, and perhaps that would be a comfort to my parents.'

Mrs Dinnage told me, 'You are reliving your memories all the time,' and I nodded.

She was too. 'I remember Christine with three of her friends popping in to see me earlier than usual the day before, on their way to school. Christine called in to return the little black dress I had lent her Mam for a funeral. I'll never forget it.'

She said: 'At the time Thomas, my son, was in the army, the medical corps. As soon as he heard about it, he was coming home. He got as far as Edwardsville and they stopped people coming in. Well, he nearly had to swear his life away, to say he was from here and lived at Aberfan, he had to say exactly where he lived and who we were. That was the only way he was allowed into the village. He came to see what he could do. There was just too many people around, you couldn't cope. It was difficult to coordinate.'

She went on: 'I remember going down the hill once that particular night. I had come home to get some tin stuff to open up to make sandwiches for the helpers. My son was coming towards me. "Mam, what have you got there?" And I told him and he said, "Oh, give them to me, Mam, I will take them to Smyrna Chapel. You go home."

'We all came with what we had to help the men and to also give clothes for them to change. They were soaking wet up there, digging all day and night. I had to peel their clothes off them and put dry ones back on the men. We did anything like that to help out. We felt helpless really.'

She added: 'For months and months after the disaster, my husband Bill used to put his finger in my mouth when I was sleeping and sweep my mouth to clean the silt out. He was doing that down the school to the children when they were fetching them out. He was sweeping his finger inside my mouth during the night. He didn't know he was doing it. He used to wake me up. He was still traumatised by it, but he never spoke about that day.'

I asked if she went to the funerals.

'No,' she said. 'I was here, with all the mothers that couldn't go into the cemetery. Some of the mothers tried desperately to go, but couldn't, they were too upset. They all came here. I made them tea and Welsh cakes. My home was open to anyone that day. Most of the mothers, you know, were wanting to go to the cemetery, but they couldn't face it, you see? They brought them here.'

Sonia said: 'You must have been one of the very few that survived.'

I showed them the last school photograph taken before the disaster and Sonia said: 'The children who stayed home, they must have thought, we could have been there. They, too, must be carrying the guilt.'

Disasters leave all sorts of emotions. Guilt, anger, sorrow. No-one's grief or loss feels the same.

Mrs Dinnage said: 'Olga and I used to go up the cemetery every night to walk along the graves and check that every light was lit and the tealights were still on. Some would blow out or burn down and we would carry candles in our pockets and make sure they were lit. We wouldn't see one without a light. This went on for a long time after the disaster.'

'I used to play there,' I said. 'Never before would I go in there, but after the disaster I was compelled to play there with David, the farm. We would often go in there to date, sit on the bench in the midst of all the children and have a cuddle and kiss or two. We were not afraid of the dark or the cemetery. Looking back now, I suppose we felt still attached to them and this gave us comfort.'

Sonia leaned forward to speak: 'I had never been up to the cemetery to the graves, even though they are right outside my door. I could never bring myself to go. I could never face it. I never got round to going until you were coming. Yesterday we both went up to the graves. It was so, so touching looking at all their ages. I saw Christine Prosser, Doug and Tina. They are all together now.'

'The cemetery is beautiful, breath-taking, an amazing memorial and tribute to them all,' Sonia added. 'I spoke to a gentleman from Crickhowell; he was there with his family from Nashville. The gentleman said that he brings people here. "It's so overwhelming. A place in history. You get a different feeling when you visit." This is the first time I went up and saw it. Since then I am very tearful. You can't change these circumstances, but it took me back to that day, when I wanted to give Christine a lift. Reliving the memories.' Sonia's voice was breaking. 'I was overwhelmed yesterday when I saw it.'

Sonia said that since our meeting on the embankment she had gone into 'Aberfan mode', going back to memories she did not think she still had.

Sonia looked at me and said: 'You are quite special. You have been able to put into words something that so easily could have over time been forgotten and people can relive it through you. It's just a true story and true stories are the best stories. You have to see it through to the end. When you put something traumatic and poignant into words people can share. It's good what you are doing. I think it is good for people to talk if they want to. If you hadn't come, I would have never gone to the cemetery.'

Another memory of one of the lost children had come to Mrs Dinnage. I could see it in her eyes.

'Avis Sullivan,' she said. 'I remember my father going out that morning. Uncle Will. He knew it was Avis when he identified her, by the beautiful clothes she had on. Her mother had lost her first child, she had a hole in the heart at six months. She lost her two children.'

Sonia said: 'I remember Olive saying that Avis had said: "Mam, don't tidy my bedroom because I am in the middle of playing in it."

'Her bedroom was right next to my bedroom so we used to knock the wall to one another. Ten years later, the bedroom was still the same as she left it. I think lots of bedrooms in Aberfan stayed the same for many years.'

Chapter Eleven

The Soldier who Cried

ABERFAN WAS A national disaster. It was also a televised disaster, one of the first. Broadcasters stood in the rubble, spoke to searchers and bereaved parents.

On hearing and seeing these news flashes, people across Wales and beyond got in their cars and brought blankets and supplies to the valley. Others were drafted in to help.

James Bullock, of the 1st Kings Royal Border Regiment, is now 72. In October 1966 the regiment was based in Devon.

When I met him to talk about the role he played at Aberfan, he started to tell me about his own challenging past. His father had been a Protestant and his mother a Catholic. 'Her father was a very staunch Catholic from southern Ireland. He disapproved of their relationship deeply but in time he gave up and allowed it to continue. When the question of marriage came up, he agreed but with conditions. My grandfather made my mother promise that if the first-born were to be a son, he would go to a Catholic school, and he was to be named James and not after my father John. I was also made to remain and live with my grandfather until my 15th birthday.'

James went on: 'My memories of that time were not good. My granddad tried in the years to come to convince me that my parents were not good, a waste of time. He hated my dad with passion and tried every trick in the book to destroy the family.'

When James turned 18, he joined the army. His uncles had served in all three main armed services. 'The stories they told me and the picture books I read made my mind up for me and it was the best move I ever took.'

By October 1966, James had been a soldier for four years and was stationed at Heathfield Camp, Honiton, Devon. The regiment was doing its normal duties, training and manoeuvres. In the hours after the disaster many were relaxing, spread around Devon in pubs and cinemas. But they were about to be pulled from their day of R & R with a jolt.

'The police were searching for us all over Devon in order to inform us to report back to barracks immediately,' James told me. 'At the time we had no idea as to what was going on or what the problem might be. [Then] we were told about the disaster and that we were to be deployed to Aberfan to assist in helping to recover the bodies of those who perished.

'It took a short while for us to prepare. Once we were ready and formed up, we all met at the square ground in a convoy of the regiment's vehicles. We then moved to Sennybridge camp, Brecon, where we were based for the duration of the operation.

'Initially, I was the driver of our medical officer and dental officer vehicle. The battalion worked from the local community centre at Aberfan.

'I also worked at the morgue. My duties varied. My job, along with my mate who has now passed away, was to clean the bodies of the children and make them presentable. I cried many times, I had to go outside and be sick. It turned my stomach, that thought has never left me.

'I was only 22 years old at the time. During my service I have seen lots of death, but nothing will ever compare with Aberfan. Aberfan should never have happened; it was a peaceful place, not a war zone.'

James told me they worked in Aberfan for about a fortnight. But that short time would have a lasting effect on him.

'For many years I battled with nightmares, drinking, depression. I received medical treatment and counselling. The memories of that disaster have remained with me for the best part of my life. I am constantly reminded of this when every time I see suffering in any form. If there is one thing that annoys me it is that there are people who knew of the impending disaster and they could have done something to prevent it happening.'

James added: 'I have experienced much in my life. I have seen poverty, deprivation, destruction, and death. You are taught about these things in school, as a child, but as an adult to see it for real sheds a completely different light on life. As a soldier my life was filled with all those things.

'The battalion are old men now and we try to retrace our steps in whatever time we have left – not that we are dying! I think sometimes we do things because we really care. Over the years many of the regiment have been back to Aberfan so that they could remember and pay their respects in their own way. For many years I had wanted to return and felt that it should be done as a group in a manner that would be treated with compassion and respect in the memory of all those who perished during that terrible disaster. We will return in 2016 to perform that duty once again.

'I think the memory of this will never die because it will be bred into the youth of the village for many years to come and through that it will never be forgotten. This is a piece of the past you can't bury and never will.

'You ask me what it means, how I feel, how my comrades feel about such matters. I can only speak for them up to a point in respect of the things we have all experienced over the years. What we did military-wise was in the line of duty, most of which was for the good and benefit of this country, but Aberfan was a period that rocked the country, if not the world. It was something that should never have happened.'

James and I first met in 2013. He said that on that first meeting he felt the joy of meeting a survivor but the sadness of seeing that I had grown up with it on my mind every day.

'You see, Gaynor; we too share in that torment, but you're more distressed because you were in that school. It's almost 50 years since I was there and I still remember it. I suppose I will forget when they put the lid on my coffin, but not for a long time to come.'

He added: 'Gaynor, there are a lot of memories that we don't want to talk about. That's nothing to be ashamed of. They're just too graphic and they will remain buried.'

Chapter Twelve

Royal Visitor

IN THE LITTLE blue school diary I kept between the ages of 12 and 14, I wrote: 'The most important visitor I had at St Tydfil's Hospital was Lord Snowdon, who sat on my hospital bed and read me a story.' Little did I imagine then, as an innocent eight year old, that he, a member of the royal family, and I, living such different lives, would one day repeat our meeting some 49 years on.

Antony Armstrong-Jones, a photographer by trade, became the Earl of Snowdon after his marriage to Princess Margaret in 1960. He took the title as a nod to his Welsh ancestry.

He was the first member of the royal family to visit Aberfan. He was 36 at the time, around the same age as many of the bereaved parents. My own parents were 33 and 30.

In our village, this pillar of society found himself to be powerless in the face of horror and devastation. Informal and compassionate in nature, eschewing all protocol, Lord Snowdon spoke to bereaved parents, weeping men and women, strong miners who just broke down and cried, offered a hug here, a quiet word there, and comforted them in silence.

He was so down-to-earth; he just had that special gift for empathy and knew just what to say or do.

He could have no idea then, or in the years to come, about how influential he would be to my journey of recovery.

One afternoon I felt compelled to research his address and contact details. I plucked up the courage and rang the number in Kensington. His PA, Lynne, answered and, as we talked, I found that she quickly connected emotionally with my story and took the time to listen. Lynne gave me Lord Snowdon's email and postal address so that I could send him a request to meet. She also stated that as Lord Snowdon wasn't feeling too well, she would also relay the message herself.

I gave my letter and email an enormous amount of thought. How would I approach writing it? Why did I feel this need to talk to him and interview him? Putting this desire into words didn't come easily but I thought I would just be me and I poured my heart out. As I wrote I felt a sense of urgency to get this done. I was also fascinated to learn that both Lord Snowdon and I shared the same birthday week, such a coincidence. He was a Piscean, compassionate, sensitive; maybe there was a sort of spiritual purpose we shared together. I knew Lord Snowdon would understand.

My letter was accompanied with a birthday card as his birthday, 7 March, was two weeks away. Oh, God, I thought, me interviewing a member of the royal family! This just wasn't heard of. The anticipation and excitement was overwhelming as I waited for his reply.

I was in the bath twiddling with my phone when I saw Lynne's email come up. My heart was racing. I was both very excited and a little shocked: a big part of me had not believed I would even get a reply. It read, 'Dear Gaynor, it was good to speak with you and thank you for your letter and card. I discussed your email and letter with Lord Snowdon who would be delighted to see you.'

I rang my sister immediately to tell her the news but told her not to tell anyone else. I didn't want it to get any further, as a private meeting between an Aberfan survivor and a member of the royal family would be bound to attract media attention. I wanted our meeting kept quiet, however much I longed to share the news. But it was really hard to contain my emotions.

I then contacted Melanie Doel. Mel was a BBC journalist who had made countless programmes about the disaster and the village. She was trusted in Aberfan and had formed close links with many of us, myself included. Mel agreed to join me for moral support and basically show me the way. She knows how to prepare for a meeting like this. Knowing our time would be limited with Lord Snowdon, we spent ages deciding on our questions.

During our journey to London, I felt very nervous. My mind and body was a whirl of emotions about how I would react to meeting him again. I was very excited, but somewhat apprehensive about how the interview would go. I'm not a professional interviewer and I don't have the distance from the subject of my story: I am part of it. I am aware of that. I am doing this precisely because *I am* part of the story.

I began to wonder if he would remember much about the events? Would I have enough time to ask my prepared questions? Would I be able to take photos and record the interview? The feelings were intense.

Mel kept reassuring me. 'Just be you,' she said, and the interview will flow. 'He will love you.'

Our coach arrived at Victoria Station at 9.30am and we jumped on the Tube.

It was such a beautiful day. The sun was shining brightly. Mel took me around some of the sights and we had time for coffee across the road from the House of Lords. God, how my feet killed, so I kept changing my shoes from high heels to flat shoes, then back to the high shoes to give me confidence for the reunion.

Our appointment with Lord Snowdon was for 11.30am and was to be at the House of Lords, or so I had assumed. I didn't even re-check or bring the email confirmation with me. A mistake! But not one that was new to me: I do it all the time. Those who know me will confirm that I can be a bit dizzy sometimes. I just hide behind the blonde hair.

We walked up to the guards at the House of Lords and I explained I was interviewing Lord Snowdon. There were surprised faces, or maybe that was just a bit of insecurity on my part. I don't think they believe me, I thought.

We were sent through the peers' entrance to the security

desk, where our photos were taken. I told the male receptionist why I was there and he looked surprised. 'Lord Snowdon hasn't been here for a long time,' he said. 'Please sit down and I shall make some enquiries.'

My heart was pounding. Did I have the right day, time, place? Mel looked like she wanted to kill me at that point.

Then when a security man came over and said, 'Lord Snowdon is not in today', the alarm bells in my head began to ring even louder. Panic! I rang Lynne from the reception, who clarified we should have gone to Lord Snowdon's home in Kensington. I had wasted time: I only had a short interview slot. Bloody hell, this is going to be a wasted journey! What had I done!

My nerves were jangling as the guards were told to order a taxi for us. They moved the barriers apart and we were quickly out into the Westminster traffic. The journey was 20 minutes but it seemed like hours with every traffic light red in our path. I was also on a budget, so I panicked about whether I had enough money to pay the taxi. What a nightmare!

Finally, we arrived. By now, my emotions were running high, my stomach felt sick with anticipation. I was sweating. I knocked the door, wondering who would answer. When it opened I immediately felt at ease. I was welcomed by Lynne, who gave Mel and I a warm welcome. We were taken into a small reception room where Lynne told us Lord Snowdon had not been very well. She then explained the protocols: we were not allowed to tape the conversation but we could take photos. This really worried me: how would I capture all the conversation? It was a case of thank God for Mel's shorthand, as we were really keen to have a true record of the meeting. Lynne said she would monitor the interview and if needed would stop the interview if Lord Snowdon didn't feel well. I just couldn't thank her enough for what she had done for me.

We were then taken to another waiting area in a small front room while Lord Snowdon prepared for us. Lynne told us the story about the red chairs we were sitting on. There were six in a row. The chairs showed off an embossed gold Prince of Wales feather crest accompanied by the motto 'Ich dien' ('I serve'). 'These are chairs from the investiture of Prince Charles when he was crowned at Caernarfon Castle on 1 July 1969,' she explained. Lord Snowdon had designed them for the ceremony and now they had been returned to their home. I was excited; it was such an honour to be sitting on them: who knew which royal bum had sat on these previously?

The door opened and in came a young lady, who I immediately recognised to be Lord Snowdon and Princess Margaret's daughter, Lady Sarah Chatto. She was petite and the image of her mother. Accompanying her was another of Lord Snowdon's children, Lady Frances von Hofmannsthal. Sarah greeted me with open arms, shook my hand and said, 'My father is looking forward to your meeting. It's a very special moment for him.' They were all so friendly and they really made me feel so special.

We were then taken upstairs to another sitting room to wait for Lord Snowdon. There, Mel and I gazed in amazement at the royal memorabilia. We were captivated by all the photos of the royal family: the Queen, Princess Diana, personal wedding photos and many of Lord Snowdon's family.

Then Lynne came in and said, 'Lord Snowdon is ready to receive you', and my heart thumped so fast. I was asked to walk up a winding iron staircase to his study. I opened the door and Lord Snowdon was sat at his desk.

My next action could easily have ended the meeting before it had started. Without giving it a second thought, I took hold of him tightly, embraced him and kissed him on the cheek. Then, grasping his hand tightly and not letting go, I said, 'This has meant so much to me, thank you, thank you, Lord Snowdon.' I took a long look at him. 'You are looking well,' I said. What possessed me to say that I don't know. It just felt the normal thing to do.

I had so wanted to meet him again and here we were. We had little in common on the face of it but we had both been touched by the awful moments of one particular day. Now, facing each other, memories came back for both of us. I felt that Lord Snowdon was clearly overwhelmed too. We just grasped hands with an overpowering intensity only he and I could understand. I called him Tony and Lord Snowdon, I hadn't given it much thought but it went down well.

He said, 'Do you speak Welsh?'

When I told him I didn't, he joked, 'Shame on you, girl!'

I said: 'If you want me to speak Welsh, I promise to take up some lessons, just for you.'

We both laughed.

I'd brought photographs to show him and I laid them out on the table. They were old black-and-white snaps and newspaper cuttings showing Pantglas Junior School and the moment Lord Snowdon climbed through the classroom window in all the chaos and debris accompanied by Lord Tonypandy.

There was also a photo of me as an eight year old in hospital with my leg injuries and some personal family photos and pictures of the cemetery and the Garden of Remembrance.

Looking at my photo in hospital he smiled gently, 'That's sweet.'

He seemed stuck for words; what could he say? 'The people of Aberfan were very, very brave,' he said quietly.

He remained silent with his thoughts for a moment. No-one knew how his visit really affected him, and I wanted to find out.

He asked me if I still lived in Aberfan and how old I was then and now.

For me, the emotional attachment felt intense. I could not stop myself from being tactile, touching his hand, arm or occasionally his knee but with the greatest of respect. His PA was present throughout. She was Welsh and the meeting truly touched her heart too. Lord Snowdon asked me to join him in a toast to the Queen. 'What would you like to drink, my dear?' I said I'd just like water. 'Have something stronger, girl!'

Lynne poured Lord Snowdon a red drink, I think this was a bloody Mary, and he asked us to raise our glasses. Lord Snowdon clinked my glass and shouted loud, 'To the Queen!'

The adrenaline was rushing but he made me feel so relaxed and excited. I felt as if we had known each other for years, but I had to get those questions in. I kept telling myself, 'Be focused! Get what you came for as time is so precious.' Melanie was keeping a note of time and would nudge me if I began to run out.

Lord Snowdon asked about my parents. 'Please give my regards to them both!'

I found the moment and the courage to ask the first question. 'Lord Snowdon, how did you first hear the news about Aberfan?'

'On the wireless,' he said. 'Then Jones the Gas rang.' Mervyn Jones was the head of the Gas Board in Cardiff, a good friend of Lord Snowdon's. 'I was at home at Kensington Palace. I remember saying, "I should be there, I am Welsh". And I just went.'

Lord Snowdon didn't consult the Palace, he just left. Every drop of Welsh blood in his veins rushed to the fore and he knew he had to go at once. He had said, 'The Welsh stick together'.

At 2am on 22 October 1966, the dignitaries waiting for Lord Snowdon in Cardiff were the Lord Lieutenant of Glamorgan Sir Cenydd Treharne, Mervyn Jones, Lord Tonypandy and the Minister of State at the Welsh Office. When he didn't appear from the first-class carriages, they scanned the platform in dismay, thinking that he had missed the train. Then, alone, without escort, he appeared carrying a suitcase which included a shovel.

In 2006 Lord Snowdon wrote in the *Western Mail*: 'I just got on a train and went straight down. I can't remember why I went second class, it just seemed the natural thing to do. I was not representing the royal family. I didn't want any fuss or bother made of me.'

As we sat opposite each other Lord Snowdon smiled at the memory. 'The officials only looked in the first-class carriage assuming I was there.'

I giggled and replied, 'You were royalty and their assumptions were wrong'.

I imagined that this was possibly the only time that a member of the royal family would travel in such a modest and economical way. Would it happen today?

Then he surprised me again, clasping hold of my left hand, holding my arm high in the air, and he began to sing: 'Wales! Wales! Join me and sing!' We laughed and sang. I couldn't stop smiling. I felt amazing, special, if only I could have filmed this once in a lifetime moment. I could see that he wanted to rejoice and re-affirm his Welsh heritage, culture and its traditions. The joy of this was written all over his face. 'You are so funny, Lord Snowdon, and have a sense of humour too,' I heard myself saying.

He was not a man for protocols. And he wasn't back in 1966. I took him back to his story.

Following a few hours of sleep in Cardiff, Jones the Gas and Lord Snowdon had travelled to Aberfan.

In his article he wrote: 'I felt at home there. I didn't feel awkward or out of touch. I was a fellow parent and so I could relate to the terrible grief those who had lost children were going through. It was horrific seeing the loss of all those innocent young children. Life seemed so unfair. I remember feeling very grateful that my own were safe and well. It was a very humbling experience. I wasn't there to take photographs either – I didn't even think about that, it would have been far too intrusive.'

I asked him how he felt when reaching Aberfan, being confronted with scenes of destruction and desolation. 'I felt very useless and unable to help,' he told me. 'All I really did was make cups of tea and coffee. I went into quite a few homes. The important thing was not to interfere. All I could do was talk to as many people as possible. I don't really remember who

I spoke to, I was too emotional. People were still in absolute shock. I just showed support and listened.'

I then explained that one of the houses he had visited was that of my best friend, Diane Fudge, whose twin sister, Daphne, lost her life that day. He had comforted Diane's parents, putting his arms around Joyce and listening quietly to Ken. Lord Snowdon went into the kitchen and made them all a cup of tea. I also explained that I have lived in that house and my daughter now lives at the same house.

Lord Snowdon was a fellow parent. His own children were almost five and two at the time. I asked if he had cried? 'I couldn't cry,' he replied. 'I thought if I cried it would just let the whole thing down.'

But it had an emotional impact on you? 'I just thought I was very, very lucky,' he said. 'I thought very much about that. Feeling grateful that mine were safe and well. It was a very humbling experience. Life seemed so unfair.'

I wanted to know if he felt angry that the disaster was preventable: it was man-made. 'I was very sad, but not angry,' he said. 'There's no point in being angry. But great, great sadness. There's no point in blaming people. Aberfan is very close to my heart and still is today even though I have never returned.'

The Queen was informed of the tragedy and was asked to come to Aberfan and share her grief with her people. Prime Minister Harold Wilson visited the village in the hours following the disaster and Prince Philip came as the Queen's advance party early on the Saturday morning. Prince Phillip quietly moved among the bereaved, offering words of condolences before returning to Buckingham Palace where it was believed that the Queen was told it would be better to wait. There has been some debate as to whether the Queen was right not to come to Aberfan immediately. Some say she should have come straight away but others point out that the numbers of security people, press and royal watchers which would have accompanied her visit would certainly have put extra pressure on the area. I asked Lord Snowdon did he feel the Queen had ever regretted her decision or the advice she was given? 'She's a very wise person and I think she would decide herself whether she should be there or just thinking of everyone,' he said. 'The Queen would have given it an enormous amount of thought, that I know.'

Princess Margaret had launched an appeal for toys for the surviving children of Aberfan. The response had been so great that the post office in Cardiff had had to put aside four disused buildings to house all the toys which had been donated.

I explained to Lord Snowdon that I had been sent a doll from the appeal. 'I remember the doll well,' I said. 'I could have any toy I wanted. A spokesperson came to my ward and asked what dream toy you would like? I chose a doll. A few days later, the biggest doll I had ever seen was brought to me. I will never forget that moment. The doll was bigger than me, with long black hair.'

Lord Snowdon smiled at my memory but the end of our time together was drawing near. Lord Snowdon's family were waiting downstairs to take him for lunch. But, before leaving, I gave Lord Snowdon my poems and explained that I had written them at the age of 14. I said that they were written in a variety of settings and told my deepest private thoughts and anxieties of the emotional trauma that affected my life after the disaster. I explained there was no proper counselling available then and, for me, writing my poetry was therapeutic and helped in my psychological recovery. I wanted to share them with him because they were part of the very meaningful memories we both had in common. Lord Snowdon glanced down and paused, reflecting on the content and titles. At last, he said: 'Gaynor, you are wonderful. The poems are very good.'

I asked him why he had agreed to meet me. 'This is very poignant to me. It is the only natural thing to do. I am delighted to have met you.'

The emotional connection to the past which had brought us together 50 years on seemed as strong as ever. Letting go would again draw me closer to moving forward.

Before we left I gave them a box of chocolates and a card with a message for Lord Snowdon's 'eyes only'.

If Lord Snowdon was ever able to make a return visit to Aberfan, he would find that his kindness was not forgotten.

Chapter Thirteen

White Coffins

IN THE DAYS after the disaster Merthyr council decided to create an educational centre in Aberfan Park. The *Merthyr Express* reported that 'it is felt that the sooner the surviving children of Aberfan are brought back to normal educational routine, the better'. The centre would become headquarters for the staff of Pantglas secondary school, as well as being a place where the officers of Merthyr education department could be available to discuss problems with parents. Three mobile classrooms would be located in the park for survivors from the junior school and I would eventually find myself in one of those.

Next to that newspaper report is a photograph taken in the devastation of Moy Road. It shows the visit of Lord Snowdon and George Thomas, the Minister of State for Wales. With them is their guide that day, John Reddy, a very well-known local businessman and politician who was then chair of further education. In those days that role meant that he also dealt with primary, secondary and grammar schools.

Having met Lord Snowdon, I wanted to talk to John about what it was like to be a local administrator having to deal with a terrible disaster and a royal visit. I met with John and showed him the photograph. I could see that it took him back straight away to the day of the disaster. As his memories unfolded, they were to give me another view of that day.

John had a shop at Galon Uchaf in Merthyr at the time and he had a phone call from a lady customer to say that something had happened at the school in Aberfan – 'a wall has collapsed'.

'Twenty minutes later I heard the news in more detail and I got in my car and drove down to Aberfan. I knew John Beale, the magnificent director of education, was in a meeting in Cardiff and I knew Stan Davies [Merthyr mayor] would be tied up in the Town Hall, so I went down to see what I could do. When I got down to Troedyrhiw I couldn't get through, so I parked in a lay-by in Merthyr Vale and walked across the river and through the fields to get across to Aberfan. It was the only way.' John hesitated. 'It was chaos in Aberfan. Everything was in turmoil. It was a scene I would have never imagined. And one of the first people I met was Dr Philip Rowlands, who was in charge of social services in Merthyr. He was with others arranging to put a temporary morgue in the school yard so that they could bring bodies out and clean them with respect and dignity. The idea was that they would bring the children out and prepare them for the other morgue at Bethania Chapel.'

John Reddy said that he then heard that John Beale was heading up from Cardiff. 'He was trying to come in on the bottom road so I walked down. They were taking the fencing down by St John's Hall so they could get machines, ambulances and trucks through. I met John Beale there. He was a superb man, a clear thinker, a brilliant director of education. Now this is important. He said: "Right, come with me, John." We went back up to the school. All the records of who would have been in school were in the registers and they were all lost in the school. The school registers would have been in each class, so no-one knew who would be in school that day and who wasn't in school. So John Beale arranged for people to go from house-to-house and enquire with parents whether their child was in school. That was the only way they could find out who was in school that day.'

John paused, remembering. 'They also wanted to know, where possible, what the child was wearing. That was vitally important because if you brought out ten, 12, 15 children, you needed that sort of help to identify them. We did not want parents having to go and see a number of children...'

All around him the digging at the school continued. 'The main thing I remember was the blowing of the whistle. Whenever they felt they got to somewhere there may be a child, they blew a whistle and everything went quiet in case the child was breathing.'

Later that evening John met with other councillors and one of the senior police officers at the scene. 'He came to see us and he said: "John, we have a big problem. We have Lord Snowdon coming, he is on his way *now* and we also have Prince Phillip. Now we have got Prince Philip in the infirmary in Merthyr visiting the sick, but protocol says that they mustn't meet and they mustn't be together, so will you accompany Lord Snowdon and take him to the houses which have been affected? We will keep you informed by police telephone exactly where Philip is so they don't meet." So I met Lord Snowdon and took him to some of the houses, the addresses I was given. And I remember this one house in particular. The lady was crying. I knew the father as he had been an apprentice with me, and he was crying too. Lord Snowdon introduced himself and he spoke very kindly, and said how much he and Princess Margaret's thoughts were with them; they shared their grief, and he passed his cigarettes around, I think they were French cigarettes. One of the ladies said: "Would you like a cup of tea?" and he said: "Only if I can help you make it." He went out into the kitchen with the lady and came back with a tray of tea, and we sat and chatted. We went to a number of houses together. He was emotional but he didn't cry.'

'Yes, John, he told me that,' I said.

'After that we, as councillors, talked and we said we must make a representation at the funerals. So we decided that six or seven of us would wait by the gates of the cemetery and attend every funeral that came in. We walked with the coffins to their graves, councillors together. That was a very moving experience, very moving, and then of course we had the mass funeral as well.'

John said: 'Let me look at that photo again. I haven't seen that since the day it was out in 1966. Would you call Anthony and Dot from upstairs?'

John's wife and son were upstairs during our conversation. I went to the foot of the stairs and called them.

When they come down, John said: 'Have I changed much?'

Anthony smiled: 'You needed to go on a diet, put it that way!'

John's wife Dot looks at the date on the newspaper: 28 October 1966. 'That's a long time ago,' she said.

I wondered what the impact of the 50th anniversary would have on John? The question brought back another memory he probably pushed to the back of his mind.

'Because the children did not have a Catholic primary school in the area, the Catholic children went to Pantglas, and I will never forget going into church and seeing six white coffins laid out. I knelt down and just said to myself, "My God".'

Chapter Fourteen

Tribunal of Inquiry

ON THE 26 October 1966, the Secretary of State for Wales, Cledwyn Hughes, appointed a tribunal to inquire into the causes of the Aberfan disaster. It was to be chaired by Sir Herbert Edmund Davies, a Welsh barrister with experience of mining law. He said the tribunal would look into what exactly happened; why did it happen?; could it have been prevented? and what could be learnt from it?

There was a great deal of tension around the tribunal. On hearing news of the disaster, Lord Robens of Woldingham, the chairman of the National Coal Board (NCB), had not gone immediately to Aberfan. He chose instead to proceed with his installation as Chancellor of the University of Surrey. But NCB sources wrongly told the Secretary of State for Wales that Lord Robens was personally directing relief work. When he reached Aberfan, Lord Robens attributed the disaster to 'natural unknown springs' beneath the tip. He told a television reporter: 'It was impossible to know that there was a spring in the heart of this tip which was turning the centre of the mountain into sludge.' Local people knew this to be incorrect, that the NCB had been tipping on top of springs which were shown on maps of the neighbourhood. The springs feature on an Ordnance Survey map of 1919 and a Geological Survey map of the area from 1959. Local village schoolboys had played at these springs; they were known.

The Attorney General imposed restrictions on media speculation about the causes of the disaster and there was a general feeling that earlier public inquiries into pit disasters had whitewashed the truth. The *Merthyr Express* wrote: 'A village has lost its children. Is not the bitterness, therefore, understandable?'

The tribunal sat for 76 days, firstly at Merthyr Tydfil College of Further Education and then, after Christmas, at the College of Food Technology and Commerce in Cardiff. At the time this was the longest inquiry of its type in British history, interviewing 136 witnesses and examining 300 exhibits.

As the tribunal progressed it became apparent that there had long been local worries over the stability of the tip, that the chairman of the NCB's claim that the spring underneath the tip had not been known about was not true, and that the coal board had no kind of tipping policy at all.

Lord Robens, the NCB chairman, appeared in the final days of the tribunal to give evidence and admitted that the coal board had been at fault. He admitted that it was known that the disaster was foreseeable by the time the inquiry began. Had this admission been made at the beginning of the inquiry, it was felt that much of what followed at the tribunal would have been unnecessary. The tribunal retired on 28 April 1967 to consider its verdict. Its report was published on 3 August 1967.

'The Aberfan Disaster is a terrifying tale of bungling ineptitude by many men charged with tasks for which they were totally unfitted, of failure to heed clear warnings, and of total lack of direction from above. Not villains but decent men, led astray by foolishness or by ignorance or by both in combination, are responsible for what happened at Aberfan.'

It concluded: 'Blame for the disaster rests upon the National Coal Board. This blame is shared (though in varying degrees) among the National Coal Board headquarters, the South Western Divisional Board, and certain individuals. The legal liability of the NCB to pay compensation of the personal injuries, fatal or otherwise, and damage to property, is incontestable and uncontested.'

The tribunal endorsed the comment of Desmond Ackner QC,

counsel for the Aberfan Parents' and Residents' Association, that Coal Board witnesses had tried to give the impression that 'the Board had no more blameworthy connection with this disaster than, say, the Gas Board'. It devoted a section of its report to 'the attitude of the National Coal Board' and of Lord Robens. It condemned both.

Merthyr Tydfil Borough Council and the National Union of Mineworkers were cleared of any blame for not following their concerns over the tip further. It was concluded that they had had little option but to accept the assurances of the NCB that all was under control. Nine individual NCB employees and officials were singled for particular criticism. However, the report made clear that it was a tale 'not of wickedness but of ignorance, ineptitude and a failure of communications'. No-one faced criminal proceedings but those named (and others cleared) had to live with the disaster on their consciences for the rest of their lives.

In the end no-one was prosecuted, dismissed, or demoted. Lord Robens's offer to resign as NCB chairman was rejected.

But there were further injustices to come.

*

The Mayor of Merthyr launched a disaster fund to aid the village and bereaved. Such was the disaster's effect on not only the British public but also people around the world, donations flooded in. Fifty thousand letters of condolence arrived in Aberfan and there were nearly 88,000 donations made. One came with the note: 'Please use this small amount in any way you wish. I was saving it up for a new coat. Oh God, I wish I had saved more. Yours sincerely, A Mother.' The people of the United States and Canada were particularly generous. *Life* magazine ran two major articles on the disaster and these had a big effect on readers. The churches of Toronto even established a social welfare centre in Aberfan.

By the time the fund closed in January 1967, it totalled

£1,606,929. The fund's final sum was approximately £1,750,000.

Financial experts saw problems early on. The fund's legal status was unclear and there were fears that any money it gave out immediately could affect future compensation claims against the NCB. However, local need was great and the provisional committee, consisting mainly of local councillors and local dignitaries, made some payments. In the village and among donors there were worries over how the rest of the money would be used. Some donors wanted the entire fund to go to the bereaved; others felt that it should benefit the wider community.

Eventually, the fund was given a firm legal footing. The fund's trust deed specified that the money was 'for the relief of all persons who have suffered as a result of the said disaster and are thereby in need' and 'for any charitable purpose' for the benefit of persons or children who lived in Aberfan and its immediate neighbourhood on the day of the disaster. Under this wide remit, the fund paid for a memorial, house repairs, holidays for villagers and a community hall.

*

As noted, the fund for Aberfan closed in January 1967 after people from all over the world had helped raise more than £1.5 million. They had donated it for the people of Aberfan. But, after the NCB and Treasury refused to accept full financial responsibility, it was revealed that the fund would have to contribute to removing the tips from around Aberfan. Initially, the government wanted £350,000. It was then reduced to £150,000.

There was outrage. But in August 1968, the Government forced the trustees of the Aberfan Disaster Fund to pay a contribution to the cost of removing the remaining NCB tips from above Aberfan. This was despite the fact that the tips were in a place where they shouldn't have been.

This demand was bitterly controversial. People wrote to ministers to complain that this was not why people had donated the money.

The campaign to get the money back would take up a great deal of my dad's time and strength.

People in Government would promise my dad and other campaigners like Gerald Tar over and over again that the £150,000 would be returned.

I wanted to ask Dad about how much this took out of him, after all he was doubly bereaved. What a massive sense of injustice he must have felt. Much to my shock and horror, Dad told me in a raised voice how he ended up in the Merthyr Vale police station over his campaigning activities and that he spent the 'bloody night in the cell'. He said, 'Oh God, I remember it well. Fred Gray came to the house and said, "Cliff, we have to do something now!" I said, "Yes, all right, leave it to me." Iris and I stayed up all night, ripping bed sheets to paint on them. I can see it now: BEREAVED AFRAID TO SAY ANYTHING ABOUT THE FUND. The next day I nailed it to two posts on the gates of the bungalow. I, Fred Gray, Billy Lucas, then campaigned.'

A local police sergeant came and arrested my dad. At the police station another officer said to him: 'You are a troublemaker, you are? You are trespassing and causing trouble in Aberfan.' My dad told him that was wrong.

'There was no sympathy from them. They were doing their job and they just didn't care what I had gone through.'

Dad was very angry as he described all this to me. He remembered saying: 'I am campaigning for our rights. That money in the fund was donated for Aberfan's residents and survivors.'

He told me: 'I was frustrated. People from all over the world were giving money, medals and coins – but not for the Government to take the money to pay for the tips.'

My dad shouted at the policeman that the money was not going to be used to clear the tips. 'I had to stay in the cell all day, night,' he stated. 'I sat in the cell. I will never forget it. They never gave me a blasted dinner either!'

According to Dad he was fuelled by anger and grief. He could not rest.

The next day word had got out to the press from Fred Gray that my dad was locked up. A reporter was on his way from Bristol. The police then informed my dad he had been demonstrating on his own ground, so they would not be charging him. 'We had better let you go then,' said one policeman, 'but I want you to sign here before you go. So that you won't cause any more trouble in Aberfan.'

'I am not signing that,' Dad told him. 'You do what you bloody well like, but there is going to be a lot of press outside when I walk through the door.'

Sitting in my parents' front room, I was totally mesmerised by the way my father told his story. I could see how fresh the anger still was in both Mam and Dad.

Dad said of the people who gave. 'Some sent their lifetime possessions. Coins, medallions.' And these were the kind people the Government was taking from.

'We were going through hell at the time, and enough was enough,' said Dad about the £150,000. 'We didn't have time to grieve properly. Fighting all the time, every day, there was constant bloody murder. Fred Gray and Harry Wilshire were up the Town Hall chasing the councillors, politicians. They gathered all the inside information they could. They heard the Government was going to take the money.'

My dad was anxious, angry, but to me and others he was a hero, a leader; his resilience and stamina just kept him going. 'How can this biggest insult in political history be allowed to happen in Aberfan? People were too much in grief to fight back.'

*

Labour's Stephen Owen Davies (always known as S.O. Davies)

was the MP for Merthyr Tydfil from before the war until 1970 when he was de-selected by the party. He then fought as an independent Labour candidate and was re-elected. As the chairman of the Aberfan Memorial Fund, Dad went to meet him. By this time they were desperate for help. 'I was fighting for the money all the time.' S.O. Davies was an outspoken man against his party's shortcomings and became very critical of the way the Labour Government dealt with compensating the Aberfan families, but by the late 1960s he was becoming sidelined by his party.

Dad also campaigned with Ted Rowlands, who became the local MP after S.O. Davies's death in 1972. They met in a chapel in Merthyr. 'In the vestry of all places,' said Dad. 'It was quiet.' Ted was to become a massive source of support for the committee. He said the situation was bad through and through. He promised my dad to see what he could do but he was not hopeful at the time. Dad told him, 'I don't mind how long it takes, Ted. I've got this far, and I will go all the way to get the money back.'

Dad then decided to take more direct action. Edgar, my dad's brother and a coal miner, lent the campaign a coal lorry off the mining yard. They filled it with muck and coal from the colliery, drove to the Welsh Office in Cardiff and tipped it over the steps. The photo caption in the newspapers was: 'The Price of the Coal: the coal that killed their children.' My dad said, 'I would be locked up for that now. I would have done anything.'

I am engrossed but, seeing my dad so ill now, I am saddened to hear how much energy and endurance he had to give to the fight. I am totally captivated by his memories, he just keeps talking, not stopping. He recalled how the anger that built over the use of the fund hit the headlines. 'I was having letters at the house from all over the world,' he said. 'Youngsters, old age, everyone. There was so much anger. They were going to remove the tip with the money. Bloody hell. One letter came from Manchester United Football Club, who had previously donated £42,000 for the people of Aberfan. They told me that they were going to claim the money back because it wasn't going to the right reason. It was all over the press.'

The pain this put my father through still disturbed my mother. 'I hope their conscience kills them,' she said.

Dad said: 'We organised a bus up to Number 10 Downing Street and the bus was full. I had travelled all over the UK gaining signatures, as far as Liverpool, campaigning to get the money back. Harry Wilshire, Fred Gray, who couldn't walk, Emanuel, Mrs Simmonds and Ted Rowlands, many more.' Dad met Prime Minister Harold Wilson. 'He said, "What can I do for you?" I just wanted to strangle him really. I gave him the petition. Wilson said, "Mr Minett. Laws cannot be broken, but in your case laws can be bent."'

It seemed that Wilson was going to look into it, put things right.

'I will never forget that moment. Ted Rowlands also told Wilson, "You will be paying this money back."'

But Dad never heard back from Harold Wilson and he had to fight on. He thought he could see it through. He was fighting the highest in the land. 'It makes you think how we have gone through it all,' said Mam.

'It wouldn't end till we had the money back,' said Dad angrily.

Mam replied: 'Never mind about the money, Cliff, I am on about how did we get through it all the fighting.'

Now I can see the strain etched onto their faces.

'Many can't stick it,' said Dad. 'The soldiers of war. The fight will either break you or kill you, that's for sure.'

Chapter Fifteen

'Were You Close
To Your Dead Child?'

THERE ARE MANY people and organisations which let Aberfan down. We got used to being a community which always had to fight for what it deserved.

One person who has always been on our side is Professor Iain McLean, the man whose research did so much to keep in the public eye the injustice we suffered and who worked to persuade the Government to return money to the fund. He has been tireless in his investigations to help the families of Aberfan.

I had been in contact with Iain for some time. In the late 1990s my nephew, Ross Manning, was studying biology at Jesus College when he saw on the Oxford University intranet website that someone at Nuffield College, named Iain MacLean, was writing a book on Aberfan. As the only Aberfan Welsh resident there at that time, Ross contacted Iain and Iain invited Ross for lunch at Nuffield. Ross also met Martin Johnes at the same time; Martin was working with Iain on the book. Ross could see that both Iain and Martin's intent was to uncover the truth and, with no hesitation, Ross offered to connect them to people he knew back home. Ross was proud that his university was contributing.

I was then contacted by Iain and Martin to help them. I first met Iain in the Aberfan Community Centre and I was immediately touched by his emotional attachment to Aberfan and his passion, personal sense of injustice and commitment to uncover the truth behind the lies.

When I started this journey I knew I needed to speak to him again. No one academic or expert knows more about the disaster than him. And he has always remained committed to our cause. It intrigued me why, as an outsider, he had taken so much interest in our story. We met at the Castle Hotel, Brecon, with the sun setting over the mountains on a Friday evening. 'The research on Aberfan is the most important research I have ever done in my career,' Iain told me, 'and it doesn't stop. So today, meeting you, is part of that process.'

I asked him to go back and explain how he became so fascinated in the first place. He said that in 1966 he was a student at Oxford University. 'We were actually on an outing that day in 1966 with some other students, taking the day off. We were at Symonds Yat which, as it happened, was as close to south Wales as I had ever been. There was something in the air, I knew something was up. A postman came rushing in and he was very distraught. I had no idea what it was about, then I got home and I saw it on the telly. I organised a fund locally for Oxford students, a reasonable amount of money, and I sent it to the disaster fund. The next thing I know, a couple of years later, the fund was being raided by Alfred Robens and Harold Wilson between them, so that gave me a personal sense of injustice which I have had with me ever since.'

*

The battle by bereaved and survivors to get help from the fund – because of the stance taken by the Charity Commission – becomes even more astounding when you consider what comes next: the decision to take money from the fund to help pay for the removal of the tips.

Villagers had wanted the tips removed for obvious reasons. They were a constant reminder of that day. Plus they were dirty, causing slurry to flood down Aberfan's streets during a storm in August 1968. A deputation from the village burst into the Welsh Office to demand their removal; they left a handful of coal slurry on the desk of George Thomas, who had become Secretary of State in April 1968. Welsh MPs backed the village and so now did George Thomas but, in light of the NCB and Treasury's refusal to pay, he sought a £250,000 contribution from local interests. This, it slowly dawned on Aberfan, meant them, or more precisely the fund set up for them. S.O. Davies, the local MP, called this 'the meanest thing I have seen in 34 years in Parliament'. Apparently seeing no alternative, the management committee met with George Thomas in August 1968 and agreed that £150,000 could be used from the fund. No-one wanted the risk of the tips staying above Aberfan. S.O. Davies resigned from the committee. In his book, Iain McLean states: 'There is no bar on a charity providing a service that would otherwise be supplied at the expense of the state and taxpayer. However, given the NCB's liability for the disaster and its previous failure to observe its own recommendations on tip safety, it is surprising that the coal board did not remove the tips on its own initiative. This was especially so since, even with assurances that the remaining tips were safe, material had continued to wash down into the village from the spoil heaps after the disaster.' Two of the remaining tips were around 100 feet high, five times higher than permitted on a hillside according to the coal board's own standards.

S.O. Davies and his successor lobbied for the money to be returned. Three Conservative governments were asked to pay it back. The appeals fell on deaf ears.

*

Then, as the 30th anniversary approached in the mid 1990s, Iain McLean realised that documents relating to the disaster would be released from the National Archives. He did some checks – because sometimes the Government withholds these documents for longer – and confirmed that many papers relating to Aberfan were due for release. 'I then went around a couple of the papers, *The Independent* and *The Observer*, and said, "This is a big story from the archives of Aberfan, could I be your reporter?" *The Independent* wasn't interested but *The Observer* was. The then editor of *The Observer*, Will Hutton, promised to give me a whole page. I spent two solid days in the archives doing nothing except Aberfan, but I suspected that the place to look for the most damning evidence would be the papers from the coal board itself, which I spent much of my time on and then it spun on from there.'

The opening of public records under the 30-year rule had revealed new information about the behaviour of the NCB, the Ministry of Power, the Welsh Office and ministers in the Wilson government, in the aftermath of the disaster, and had given Iain ammunition to write a number of newspaper features and an academic article about the failure of the Wilson government to hold anybody responsible for the disaster. Iain revealed that Lord Robens never intended his resignation to be accepted and that ministers let him stay because they thought he was the only man who could manage the decline of the coal industry.

Iain McLean then went on to write a book called *Aberfan: Government and Disasters* (with Martin Johnes) which came out in 2000. 'I retained an interest in it ever since,' he said.

So when did he actually first visit Aberfan?

'I deliberately didn't visit Aberfan [in the years after the disaster]. I saw it off the old main road. In the early years I thought this is not the place for casual tourists. I didn't go to Aberfan myself until Martin and I were working on the research project. I felt that after 30 years this is still a place that has a lot of issues... It is not for me to intrude unless I am invited.'

I'm proud to say mine was the first voice Iain quoted in his influential article, 'On moles and the habits of birds: the unpolitics of Aberfan', which he sent to Ron Davies, Secretary of State for Wales, immediately after the election of the Labour government in May 1997. In his covering letter, Iain asked the new government to consider repaying the £150,000 to the Aberfan Memorial Trust, which maintains the cemetery, where most of the victims are buried, and the memorial gardens on the site of Pantglas Junior School.

I asked Iain if he thought the Labour Government in 1997 was unnerved by the release of the new information and evidence?

'Absolutely. So the first thing I did after the change of Government was to write to Ron Davies and I said, "I am not going to be the only one saying this to you, but here is something I really think you should do: give Aberfan their money back".'

On 31 July 1997, Mr Davies did exactly that, citing Iain as one of the people who had led him to that decision.

'I remember a call came through from Ron Davies's private office in Cardiff saying the Secretary of State is about to make an announcement saying he is giving the money back, we thought we would give you some warning as we are going to mention you as one of the people responsible for this. Then there was Radio Wales, where you came on; both you and I were being interviewed the same time, remember that day?'

I did. I was working in an office in Cardiff when my father rang me to tell me the news. I'd broken down in floods of tears – tears of happiness for my father who had fought for years to get the money back, going through all those years of trauma, bitterness, hatred; and of great sadness for those who had fought with my dad and died early with grief so they were not there to see this victory. I was so emotional and upset that I had to be sent home from work. 'Mixed emotions for us all,' I told Iain.

*

I asked Iain if he felt that any of the Government's behaviour had been unlawful? 'There was a lot that was just dreadful but not unlawful,' he said, 'but what I think was clearly unlawful was taking the money to remove the tips, because your fund is a charity. There are a number of things that a charity may be legally used for and the removing of the public coal board tips wasn't one of them. Furthermore, there was no question that the coal board had liable intent, as a lawyer would say, strict liability for all the consequences of the disaster, and they were clearly liable for removing all the other tips. What we know is that Alf Robens's game was to make sympathetic noises when he was in Merthyr, saying that the tips ought to be removed, while in back channels in Westminster they were saying the coal board wasn't going to pay for the tips.'

Iain says that he felt the payment was unlawful and that we should have been protected by the Charity Commission. 'They have now, since our book came out, held their hands up and said they were wrong.'

Lord Robens lived for 30 years after Aberfan but he never apologised for his part in it. When George Thomas passed on a plea from worshippers at Bethania Chapel to have the building demolished and rebuilt (they found it unbearable to worship on pews that had been used as mortuary slabs), Lord Robens rejected it. The disaster fund eventually paid for a replacement building.

I asked Iain how it was that no-one was fired, or even demoted, in the wake of the disaster.

Iain shrugged. 'That's all in the distant past now,' he said.

In 2007 the Welsh Assembly Government went further than Westminster and paid £2m to the Aberfan fund to cover for the real losses made because of the money taken to clear the tips. The money would be used to safeguard the future of the memorial garden and cemetery. 'The second wave of giving the money back, which was done by the assembly government, at least brought you – the trustees – back to the financial position you should have always been in. That wasn't done by Ron's transfer because he only gave the same nominal pounds back. So it was the second transfer, done by the Welsh Government, which returned the amount which represented what had really been taken. I don't claim any responsibility for that, except about making a general fuss about saying the job isn't done yet.'

A major issue was why no-one was ever prosecuted over the disaster. Could Iain make sense of that?

'It is a bit of a puzzle because the idea of prosecution for corporate manslaughter was not unheard of. There had been a case a year before, involving a contractor demolishing an old railway bridge.'

I had read about this case in Iain's book. A welder-burner had died when the bridge had collapsed into the river Wye. His employers were prosecuted for unlawful killing. It failed because it could not be shown that the instruction which had led to the collapse had come from the managing director. Therefore, the company could not be liable.

I said that for many Aberfan parents, families and survivors, a successful prosecution for corporate manslaughter would have given them closure.

Iain said: 'In retrospect, in my opinion, it is obvious that the coal board corporately and its director of operations should have been prosecuted. Probably not Robens, in fairness, because he was the chairman of the board. It should have been a criminal prosecution for corporate manslaughter.'

I sighed and said I still felt that there had been no real justice for the people of Aberfan. I felt Iain might be the one to tell me if there ever would be.

He thought for a moment, and said: 'You can't bring children back.'

*

Iain's answer knocked me back. Maybe a search for justice or compensation or anything tangible can only ever be meaningless when the issue at the heart of a disaster like Aberfan is the enormous loss of life. We all know nothing can ever fill that hole in a family's life which has been left by the death of a child or loved one.

In his book Iain McLean explained that because the original appeal for money for Aberfan had been made without 'clear objectives' there was no single and specific intention behind each donor's gift and tracing tens of thousands of people to ask if they wanted their gift to have a specific purpose would have been completely impractical.

The trust deed for the fund was set up the month after the disaster and stated the money was for 'the relief of all persons who have suffered as a result of the said disaster and thereby in need'. It also stated it could be used for 'any charitable purpose for the benefit of persons who were inhabitants of Aberfan and its immediate neighbourhood' on 21 October 1966. A memorial was built early using the fund but people were now in need of the money. After much discussion with the management committee, the Charity Commission eventually advised that the fund could pay £500 to each set of bereaved parents because of the 'devastating effect of the disaster upon all the people concerned and the intense and mounting state of emotion which had resulted from it'. Both the parents and the management committee demanded a higher payment, as an acknowledgement that the money had been donated to the fund in sympathy for the deaths of the children. The committee argued that a payment of £5,000 would relieve mental stress and strain, allowing those who wanted to move away to do so and those who wanted to stay to make a new start. The Commission conceded that the payments would be permissible but 'before any payment was made each case should be reviewed to ascertain whether the parents had been close to their children and were thus likely to be suffering mentally'.

That piece of Charity Commission advice was ignored. The Aberfan members of the committee resigned so that they could be considered as beneficiaries. A flat rate payment of £5,000 was made to all bereaved parents. But then there were the children who had not suffered physical injury but were suffering mentally. The committee wished to support them but the Charity Commission said it could not. Committee chairman Stanley Davies said he regretted that the fund had been made into a charity. Eventually, one-off grants were made to physically and known psychologically injured children, although many others had not been assessed by a psychiatrist. As Iain has noted, 'Charity law seemed designed to protect only its own antique structure. Certainly, it protected neither the donors to, nor the beneficiaries of, the Aberfan Disaster Fund.'

So if there is no justice, if there can be no true compensation, what does society learn from Aberfan?

I asked Iain how he thought Aberfan now sits in the public consciousness and what effect it had on our country, even the world.

'It's the one that everybody remembers, if one was at a certain age, and of course younger people come across it for the first time. So it is very easy talking to people my age where you can talk about bits that are important, but talking to those younger people, as I do from time to time, I know which are the bits that are going to make them gasp: it's the bit about moving the tips; it's the bit about obscuring for the 70 days of the tribunal what the coal board knew from day one, which was the NCB had been tipping on the springs which were shown clearly on the Ordnance Survey map. It's also the one about "Were you close to your dead child?" Martin found that statement in the trustee's charity report.'

I tell Iain that when I read that for the first time, I just felt sick to my stomach. Reading insensitive and ridiculous statements like that always make me feel the anger, torment, and stress that my parents suffered.

*

Iain McLean said there were many practical lessons that had now been learnt from the disaster. 'Starting with the easy ones,' he said. 'It took until Aberfan for the mining industry to know that tipping on slopes was a silly idea – the NCB had been doing it for 100 years or more. Now all has been

The teachers who survived: Howell Williams, Hettie Taylor, Mair Morgan, Renee Williams.
Howell Williams

After the disaster, a school trip to the Isle of Man: survivors Dawn Andrews, me, Susan Maybanks, Christine Jenkins, Janet Jones.
Howell Williams

The clearing up operation continued for years as this 1968 photograph shows.
Trevor Emanuel

The tips being flattened, 28 June 1968.
Trevor Emanuel

Photographs taken in 1969.

Trevor Emanuel

Draining the tip, 1970. The photograph shows the pressure of water from the tips.
Trevor Emanuel

The landscaping work continues in 1971, replacing the old Black Bridge.
Trevor Emanuel

I was carnival queen twice in 1974. Here I'm being crowned Aberfan and Merthyr Vale carnival queen by *Crossroads* actress Meg Richardson.

And here I'm being crowned Bryntaf's carnival queen.

The fun of the carnival on our street in Bryntaf. Mam in her curlers, with some of her friends: Betty Carpenter (left) and Janet and Margaret Payne to the right.

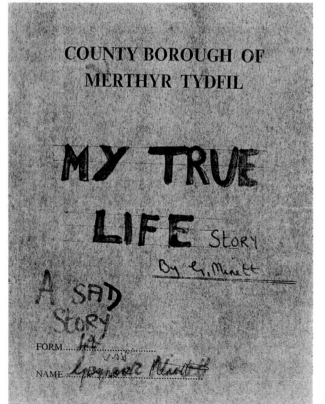

COUNTY BOROUGH OF
MERTHYR TYDFIL

MY TRUE

LIFE STORY
By G. Minett

A SAD
STORY
FORM 4
NAME Georgette Minett

It was in 1974 that I started keeping my journal.

National Coal Board officials attend a meeting at Aberfan and Merthyr Vale Colliery.
David Schewit

The Aberfan and Merthyr Vale Colliery in 1990, around the time that it closed.
Trevor Emanuel

Dad campaigning outside the family home, Carlyn, for the return of the disaster fund money, with Billy Lucas and Fred Gray.
Life Magazine

Former Welsh Secretary Ron Davies and Dad (left) share a moment in the Memorial Garden after Ron organised that the money taken from the Aberfan disaster fund was repaid by the Government in 1997. The fight was now over.
Media Wales

The Queen opens a new community centre in Aberfan on 16 May 1997. Here she is speaking to me and the other survivors.

Media Wales

Dad escorts the Queen and Ron Davies, the Secretary of State for Wales, through the Memorial Garden, with other dignitaries, including the Duke of Edinburgh, walking behind.

Media Wales

Parents and sisters. Back row, L–R: Michele, Sian, me, Belinda, with Mam and Dad seated.

Dad and Mam on his 80th birthday, 3 August 2012.

Dad enjoying a pint with his grandsons. L–R: Barrie, Dad, Ross, James, Ben.

Dad and Mam with their grandchildren. Back row, L–R: Ross, Dad, Mam, Barrie, Cassie. Front row, L–R: Ben, James.

My youngest grandson, Aaron.

My grandsons, Oliver (left) and Jaxon.

My granddaughters, Megan (left) and Lily.

Sharing memories with Lord Snowdon at his home in Kensington.
Melanie Doel

With Ron Davies, who made it possible for the money taken from the Aberfan fund to be repaid.
Melanie Doel

The view across the valley to Aberfan today.
Greg Lewis

The peace of the Memorial Garden on the site of our school in Moy Road.
Greg Lewis

To those we love and miss so very much: the plaque in the Memorial Garden.

Greg Lewis

The children's graves on the hill above Aberfan.

Greg Lewis

stopped and that's all gone now and there are no more. As far as holding people responsible, we have a much better culture now than they did then. The agency that failed was the mining inspectorate, which didn't have a tipping policy. Take the coal board and mining inspectorate separately: the coal board, all its failings were in the Edmund Davies report; the mining and quarries inspectorate were the guys who were supposed to be checking on the coal board. Look at aerial photographs in the tribunal report – aerial photographs taken before 1966 – and you can clearly see the tips bulging. You don't have to be an expert in mining.'

In his book, Iain picks apart the National Coal Board's initial argument that the slide was unforeseeable and that it was caused only by geological conditions. He points out that not only were the springs over which the waste was tipped on maps, but that there should have been plenty of knowledge to show what could potentially happen to the tips: there had been a large tip slide down the road in 1939 and the Aberfan tips themselves had slid in 1944 and 1963. As he and Martin Johnes noted: 'Tip slides were neither unknown nor unforeseeable.'

Iain told me: 'There wasn't as much made of that failure as there should have been. They should have spotted it, from the slide in 1963. They should have spotted it from the slide in 1939.' In fact, as Iain has also written, there can hardly be a clearer case of a disaster waiting to happen.'

Iain added: 'Certainly, regulators now are much better than they were then. When things go terribly wrong it has got much better, for example the Clapham railway disaster. We see it every day. The various regulators/Health & Safety/rail accidents people are more sensitive and they are less likely to accept the Lord Robens's technique of trying to blame it all on others.'

I said I am always interested when I see the 1974 Health and Safety at Work Act mentioned as it was derived from the failings of Aberfan.

'Yes,' Iain said, 'but who did they get to chair their committee? Robens! That's another gasp for our youngsters. Martin and I spoke to some of the civil servants who had been involved in that and no-one saw anything odd. Robens was this big beast roaming around, who their employees were terrified of. He wanted to be prime minister. People were scared of him. Now I understand why, he was a big bully. When the coal board came to an end, when the politicians wanted to put this potentially dangerous guy where he could be safe, they put him in charge of health and safety! Now, all right, it's fair to say that work place safety in the collieries themselves had improved a lot in his time and that's a fair comment; nevertheless I think he was the most inappropriate person in the country to chair the health and safety committee for the Health and Safety at Work Act. Didn't people have a sense of irony, a sense of shame?'

I asked Iain who responded well to the disaster.

'The tribunal did well,' he said. 'The Edmund Davies Tribunal was a great piece of work. And I think Harold Wilson did well. He wrote on his copy of the Tribunal report "THIS IS DEVASTATING". I think the good guys include George Thomas. I know there are a lot of questions around his reputation now, but he played a good role in the aftermath of the disaster and also Ron Davies and you have to include the Welsh Assembly Government now.'

Should the successors of the NCB hold the liability for the disaster?

'If the coal industry was still nationalised and there was a body you could sue... but there isn't and you couldn't. The coal industry is no longer around today so it would have to be political. You know that the tribunal named nine individuals of whom one should not have been named. He seemed to be a very honourable person and he was the colliery engineer. In the tribunal he said, "I cocked up", but I don't actually think he did. He was put in a position where he knew nothing about soil mechanics but he took the view that he was the mining engineer and he thought it was his fault. But he had no training. Out of the nine, he was the only one who owned up. I feel he should never have been among the nine... I am sure he has passed away now. You look at the various "high-ups" and you don't get many of them confessing...'

*

After the tribunal, a single inquest was resumed. It lasted only four minutes. The coroner stated that: 'It is not for me to decide any question of civil liability. As far as criminal liability is concerned, the whole matter has been investigated very thoroughly. The tribunal assessment is tantamount to a finding of accidental death in each case.'

I told Iain that until I read this I'd thought the tribunal was about the facts and not the blame. Iain said: 'The coroner's statement was also not a happy story. I only have the press reports, but I don't know if the coroner's verdict is still open, or a coroner's court could now overturn. Being overturned, that could open all sorts of ghosts, so it wouldn't be for me to recommend.'

In his writings, Iain has noted: '[The coroner's] interpretation is perhaps questionable given the force with which the [tribunal] report had condemned the NCB. Accidental death was certainly not how many people in Aberfan felt about their losses and the press reported that some parents felt the verdict should have been manslaughter. But that was not how the legal system worked.'

My mind goes to what I have read from the anger shown at the opening of the inquest a week after the disaster. One father had said then: 'I want it recorded: "Buried alive by the National Coal Board." That is what I want to see on the record. That is the feeling of those present. Those are the words we want to go on the certificate.'

Despite the harsh words of the tribunal, its report was effectively the end of the matter. No convictions, sackings, not even a demotion. With a sigh, I asked Iain how he would describe how the victims of Aberfan had been treated down the years?

'Badly,' he said flatly. 'No-one has been successful in getting compensation for Post-traumatic Stress Disorder (PTSD) which is probably still here after all these years. It took 30 years after the disaster before the fund got its money back, then another few years before the second batch of money came back. It just goes on and on.'

Chapter Sixteen

A Minister Makes Amends

Listening to Iain McLean's assessment of the story of Aberfan, I wanted to meet another of the 'good guys'. I asked Ron Davies if he would meet me and I was invited to his home. It was obviously important for me to see him, because his decision to pay the money back had made such a massive impact on my father.

But I also wondered why he felt so close to this story of Aberfan. 'Because, I guess, Aberfan is a name that will resonate with anybody who is familiar with it and I was touched by it personally and I have had the opportunity to do something in a more public role. Therefore, you know, in a sense I am involved, albeit tangentially, in the Aberfan story.'

Like many, Ron remembered where he was when reports of the disaster came through.

'I was a student at the time in Portsmouth and I can remember reading the headlines in, I think, the *Portsmouth Evening News*, not that I was an avid newspaper reader, but I remember looking at the evening paper and the headline was "Children die in tip horror" or something like that and I assumed it was a local story – just a couple of kids maybe tragically killed locally – not realising for one moment the full horror behind all. It all became apparent during the hours that followed. And for me being away from south Wales, my background having been in south Wales, it was just heart-wrenching.

'I had actually worked in the coal industry before I went to college. I had worked at the coke ovens in Bedwas and, as such, was part of the coal work culture. Miners drank with miners, played rugby with miners, and I came from a mining valley, and it was very, very personal to me and I remember the sort of feeling of desolation and angst, I suppose, being so far away and being very conscious of what was happening.'

Ron said that when he began to work in south Wales he would often visit the Valleys.

'I would travel up and down the Valleys and I guess my first engagement was in about 1970,' he said. 'I was working for the Workers' Education Association and there was a group working out of Aberfan under the Call to the Valleys programme. I sort of got involved with the community then. I was up and down and in and out of Aberfan and began to know people and understand a bit more about Aberfan, and also the place of Aberfan in the context of what was happening in the mining valleys.'

'How did it make you feel when you visited?'

'I was just overwhelmed really, just looking at the site having seen it on television. Because I think most people's memories were framed by TV and newspapers and seeing photos and so on – they were invariably close up and grainy, and you didn't really get a proper sense of it.

'I can remember the first time I went into Aberfan, going down off the old A470 and down the side of the hill there and across, and it's just difficult to describe. How can a place like that, which could have been where I was born or the village next to it, be devastated in such an awful event? I had a feeling of just desolation and compassion for the people who had been blighted by it.'

'Still today?'

'Yes, absolutely.'

I asked Ron what his thoughts were when he heard that the then Labour Government was forcing the Aberfan fund to contribute to the removal of the tip.

'I just couldn't believe it at the time,' he said. 'I had actually been active in the 1966 General Election, working for the Labour

Party as a door-knocker and leaflet distributor, and what was being done subsequently in 1967 when I came back to Wales was incomprehensible. I just couldn't understand it. Even from a party political perspective it was crass. It was indefensible, there was no reason for it. And I just was embarrassed and ashamed about what was being done in the name of the party I supported.'

S.O. Davies had been targeted by people within the Labour Party who had wanted him replaced. This had made it difficult for him to campaign for us. Ron said that the internal power struggle meant that Labour in Merthyr was 'in turmoil'. This contributed to the continuing injustice to the people of Aberfan.

'People would talk about Labour and say how on earth can you support Labour – people who had put their name to this? It was theft, it was using money for purposes it was not intended for. It represented the generosity and the humanity of people right the way across the world and here you had a Government, for no good reason, deciding to appropriate some of it, and socialist parties, to my book, Labour Governments, don't do that sort of thing, so that was very much clouding the whole of Merthyr politics.'

'Do you feel the residents were bullied into giving the money?'

'I wasn't close enough to it. It was clear that there was widespread opposition and hostility to the decision but it's worth reflecting that the Government then was a different Government to today. Then Members of Parliament were aloof, respected, far-off individuals. Government ministers even more so, so I guess if people were told this was going to happen they accepted it. If it happened today, in the era of mass communications and Facebook and Twitter and all the rest of it, people would have been out on the streets, but that wasn't the case then.'

'From your point of view you saw this as illegal or at least immoral?'

'Absolutely, I don't know about the legality of it but it was certainly immoral. It was crass, it was insensitive, it was stupid – at best it was stupid because it was so offensive. It was flying in the face of any decent emotion – indefensible!'

*

Tony Blair's Labour Party enjoyed a landslide election victory in 1997 and Ron Davies was appointed Secretary State for Wales. One of his first acts was to return the £150,000.

'Yes, it was pretty straightforward really,' he said. 'Ted Rowlands, who was a pal of mine, we were close in political terms, was someone with whom you could have a fairly frank discussion and we represented the same party. I can recall he contacted me. I think he sought me out and so it wasn't a coincidence – a casual discussion – he sought me out when I was Shadow Secretary of State [for Wales], I think it must have been 1994/95 when I was bedding into the job and we were starting to work towards the election and he contacted me and said he wanted to see the money repaid and I made it clear I was willing. He then told me that he had had a discussion with John Smith prior to the 1992 election and that he had had an agreement with John Smith and would I sign up to pledge to honour that agreement in the event of us winning in 1997 and maybe my being Secretary of State for Wales? I said absolutely, no doubt about it. It was a conversation and I can't remember if we actually physically shook hands but it was a done deal.'

'I know that you said the slate should have been wiped clean from 1968?'

'Yes, it was a bit of personal stuff in there as well, as I'd regarded George Thomas as the sort of person I stood for and I believe it would have been a personal decision of his... I believe honestly that had I been in that position I wouldn't have stayed in the cabinet for two seconds if I had been expected to make that decision, and I think if George Thomas had wanted to have stood his ground, I believe he could have stood his ground.'

I asked Ron to clarify this: 'You felt it was his personal decision to take the money from the fund?'

'Yes. I am not sure how much pressure would have been put on him but at the end of the day he would have said yes, and if he had said no – he could have said no – and if he had said, "No, I will not be part of a Government that does this", it would not have been done. So he was, at best, complicit in this, if not the architect.'

I asked Ron if he received any resistance from the Blair government when he decided to pay the fund back?

'No, not at all, not at all. I was making the point that, as far as I was concerned, it was an unbelievable opportunity for me and you don't get many opportunities in politics to right wrongs. But for me it was an unbelievable opportunity for me to right a wrong, which I could influence because it was within my remit. It was my budget and, in a sense, I was in exactly the same position – in reverse, as George Thomas decided he wasn't going to do it. I decided I was going to do it.

'I am not sure if you know how politics, how government works. You sort of sit down with your senior officials, you look at issues, and I made it clear it was pretty high on my agenda that this was something we needed to do and I also made it clear to my officials they needed to get hold of the Treasury and get the approval because at that point it was pre-devolution. I can remember my private secretary coming in and saying, almost casually, the Treasury have agreed to repay the Aberfan money and that was it. There was no great battle.'

Ron told me he felt a 'sense of inner relief' when the Treasury accepted his request. But there was a major issue with the repayment. As I mentioned earlier the fund's final sum was approximately £1,750,000 (worth about £17.5m in 1997), and the £150,000 the Government had taken in the 1960s was worth many times more by 1997 – but it was only that sum that was paid back. No interest. No adjustment for inflation. I asked Ron why that was?

'I questioned it time and time again and was questioned on it,' he said. 'And I questioned people like Ted and people involved in the fund and I was told this is what had been asked for and that was good enough for me. It wasn't for me to grandstand and say, "No, you have got to have more". I was honouring an agreement and I was told the request was the capital sum and I just wanted to honour that agreement.'

Maybe if the community had asked for more it would have been a different story.

'I can remember I was very uncomfortable when I went to Aberfan subsequently,' Ron said. 'It may have been to the launch of one of the books but I was at one function and people were sort of having a go at me and saying why haven't you paid the interest back and I was a bit – I didn't expect people to be grateful, far from it – but I was a bit taken aback, a bit surprised, because I wasn't aware that it was an issue. I made enquiries and I was told that was the deal, that was what people were asking for. I hope you understand that is my position: that it would have been presumptuous, it would have been arrogant, if I had said, "You know, I am going to override this. I am going to decide to do something they haven't asked for." I wouldn't have been comfortable with that. I was trying to do the honourable thing and as far as I was concerned, I was honouring an agreement and that agreement was for the capital sum.'

When the official announcement was made that the money was being returned, Ron paid a touching tribute to the people of Aberfan. My dad was chair of the memorial committee at the time. I showed Ron some photographs of him and of a newspaper with the headline: 'Tears flow in the rain.' I told him: 'I watched you with my dad from the garden. It was emotional, it was just unbelievable, because my dad was crying and you were crying...'

Ron became very tearful at the memory. 'There's not much you can say is there,' he said.

I told him: 'It was the end of the line. The money would put it to rest for me. I had been going round as far as Pembroke collecting money from companies who had raised money for the fund and taking it to the committee to keep the memorial going because it was crumbling, and when we got that back it was emotional.' Residents had campaigned fiercely and tirelessly to try to get the money back. Dad remembers trying three times with the Conservatives and was turned down, but he never gave up hope.

Sadly many parents and survivors did not live to see the money returned. Maybe some died of a broken heart. 'Lots of people died young,' I said to Ron. 'Parents died young, you know. What were your feelings? Obviously, you are lost for words.'

'To be honest with you I was drawing strength as much as giving support because it was hugely difficult for me, not difficult, but emotionally difficult for me, and just to be in the company of someone who had been hurt far more than I had ever been was emotionally demanding. And I just wanted to give some sense of comradeship, some sense of personal support and, at the same time, draw some strength because there are times like that that people do need a bit of inner contact, to hold each other. I don't think your father said much and I don't think I said much either. In instances like that you don't have to say much do you?'

*

Ted Rowlands was at the House of Commons during the 1997 elections. Dad remembered him phoning him at lunch-time. 'Cliff, great news. Labour got in and they have told me Aberfan is going to have their money back.'

Dad's emotion was too much to bear. Exhilaration, empowerment, great relief. But also a feeling of great sadness too, as many of the bereaved families who fought for justice had sadly passed away. They were not there to share the news that day.

I remember that day well too. I was at work in Cardiff when I took the call from my dad. I can't explain in words how I felt, not only for me, but for him. I knew what this news meant to him. I couldn't cry or laugh, I was just numbed at first. Then, later, I remember breaking down, and I had to be sent home from work.

Two hours after Ted's phone call, Dad took a call from the Charity Commission ringing to apologise for their part in the insult to Aberfan. It still makes both my parents very angry. Dad said an apology was not enough from them. 'These organisations were accountable. They should have been brought to book. It didn't matter how much time had gone by. They were accountable.'

After the news came out my parents received letters saying they should have stuck out for more. 'We were just glad for what we were having back,' said Dad. 'We had fought long enough.'

Thirty-one years it took my dad to get the money back. 'It eats you, it's like a cancer,' he said.

*

As I met the people on my journey, I think they understood that my life had been one of struggle since Aberfan. That the anxiety caused by that day is still very much present today, that it can grab hold of me and halt me in my tracks when I least expect it. I know there is no sense in feeling bitter as it does not allow you ever to move on, but I have always thought that all survivors were never fully compensated for what they had gone through, especially after the Government had repaid its debt to Aberfan all these years on. There has been no personal successful claim made against the Government or the National Coal Board, nor has there been any Government apology.

When I met Ron Davies I asked him if he thought it appropriate for a current government to atone for the mistakes of their predecessor. No-one had ever tried to sue the Government or made a claim of PTSD liability.

'That's a decision, a judgment, of a lawyer and I am not a lawyer,' he said. 'I think I can try to answer your question: Governments have demonstrated that they have a moral responsibility to right wrongs and that's why you frequently get governments making apologies, issuing posthumous pardons in the case of some criminal events, in the case of execution when some atonement has been given. But I think what governments are doing is saying those decisions wouldn't have been taken today in the light of today's values and today's standards. I don't think you can draw from that that you would have a successful case for compensation, because the rules relating to compensation are quite different.'

He added: 'I think it's right and proper for governments to express a view and I guess in this case the Government of which I was a member expressed a view by giving money back: what greater apology can you have than that, than saying this shouldn't have happened, you know, we are undoing it? Whether you then go down the road from that in saying there must be personal recompense, I don't know. I don't think so, because where do you then stop? Every single incident? Every case involving road accidents or whatever? So I don't think you can go down that road. I think it's legitimate for governments to say the wrong decision was taken and we apologise. That's legitimate.'

Although the money came back I didn't remember any one official ever saying 'Sorry'. Did Ron?

'No, I don't recall that. But I am absolutely clear, even now after the passage of so much time, that if anybody had said when I was there, "Is this an apology?", I would have said, "Yes, it is an apology and, if you want it clearer, I would say this is a Government apologising for what happened and for taking the money". So it was a case of deeds not words. The question of a formal apology, I could never recall coming before me. But what more could we do? Something has happened and we are undoing it.'

'Sometimes I still long to hear an official apologise,' I said.

'It could happen,' Ron told me. 'But it's so difficult to look back at the history. If I had said that this was an apology on behalf of the current Government on behalf of a previous Government, I don't think that would have got any profile at all. I might be entirely wrong but the real issue here was about a cheque for £150,000 and the raw emotion around that.'

Chapter Seventeen

Her Fingernails

ONLY ONE SURVIVOR of the Aberfan disaster has ever tried to battle the Government and courts to win compensation for post-traumatic stress. Chris Crocker had just moved up to Pantglas Secondary School at the time of the tragedy. She had blocked the tragedy out of her mind until her own daughter reached ten, when she had a nervous breakdown.

The National Coal Board awarded £500 to the families of each dead child. (This was entirely separate from the money allowed from the fund.) The NCB described this as 'a good offer'. It was little more than the amount paid out per farm animal. This was a payment that should have been as much about justice as finance. It was about recognition of a life lost; recognition that the loss of the bereaved had been acknowledged. At Aberfan, that money was not enough to say the NCB recognised our loss.

In 1978, this was the subject of the Pearson Commission but that recommended no change in this state of affairs. Lord Robens gave evidence to the Pearson Commission, while the Aberfan Parents' and Residents' Association did not.

Mam said to me once: 'How can they put a price on a child's life? How do they know if that child, or Carl or Marylyn, would have grown up and been a scientist or teacher?'

There was anger in our family about how much was paid. 'It makes my blood boil still now,' Dad said.

During the 1990s, Chris Crocker hired campaigning solicitor Glyn Maddocks, an expert in miscarriage of justice cases, to help her. The case was lost because the judge ruled too much time had passed since the tragedy. Glyn Maddocks said, at the time, that it was not too late for the Government to review that rule to help the victims of Aberfan. He told the *Western Mail*: 'The £500 given to parents was a scandalous amount – it was cruel, derisory and insulting and it would have been better if they had given nothing. As for the surviving children, they got nothing from the Government. Many of their lives have been ruined – now, as adults, they are suffering from post-traumatic stress, nervous breakdowns, and are struggling to get on with their lives. They deserve something, and £50,000 would be a good starting point. The Government let Aberfan down. The village lost a generation. Unfortunately, the Government wants closure on this. It doesn't want to rake up the past and so is unlikely to push it.'

In the same article Chris Crocker said that, although the claim failed, she believed she won in her aim – to stand up to British Coal. 'I wanted to show them what they had done to the children of Aberfan. I stood up in front of them and, in my mind, I won.'

*

I thought about claiming for PTSD myself. I felt that everybody would recognise that I had suffered terribly. Not only the loss of two siblings, but many feelings of anxiety. So, in 2003, I plucked up the courage up to meet with a solicitor. He quickly wanted to get a barrister involved. But I had no legal aid and my bill escalated very quickly. I panicked; I just did not have the money to pay the fees, so I withdrew my claim. This upset me enormously as I felt I was leaving the Government to walk away scot-free.

This experience left me wanting to meet with Glyn and Chris.

*

It was a hot, sunny day as I drove over the hills of the Brecon Beacons National Park towards the small town of Crickhowell.

As I arrived I thought about how this seemed to be a lovely place to live. There was a sense of stopping time here, which was what I wanted to do with Glyn: take him back to the court case, ask him how he felt about it being unsuccessful and to ask him how he felt Chris and other victims of Aberfan had been treated.

Glyn's office was in the heart of the town. He greeted me in a reception area and we went into his room where he had a box full of legal files from the case. Digging out the file had obviously brought his mind right back to what an injustice Aberfan had been. The files even included a copy of the inquiry. 'What a cover up this all was,' he said.

I asked Glyn about the case.

He said: 'She arrived on the scene a few minutes after it all happened. She was then digging around for all of her friends because she had only just moved to the senior school in the September. Not only had the coal tip buried all of her friends and her classmates, what happened then was that she buried her head in the mess. For many, many years, the trauma of what had happened affected her but it was only some 20 or 30 years later that it all came out, and she had a breakdown. She then displayed the most clear signs of post-traumatic stress disorder that the psychiatrist treating her had ever seen. And he actually videoed her experience for teaching purposes. It was so clear: on the bed in the hospital she was digging at the mattress, reliving those memories from when she was 11. It affected her life in a way that she did not know; it was affecting her all the way through her teens and her 20s and 30s. It was completely submerged in her subconscious, head and mind, and only came out later.'

He said the case was defeated because there is a limitation on compensation claims like these: you only have a three-year period from when you turn 18 to make a claim. 'She could have made her claim anytime up to 21, but during that time she didn't know, as this was all under the surface.'

I asked Glyn how he felt when the case was lost. 'I was very,

very upset. I felt that justice had not been done, and that really she had suffered. The courts should have been more willing to consider her position and the fact she was not able to make the claim earlier. I don't think the right decision was made. I feel that Christine has been in the same position as you. I am not saying money would have helped her, but the fact is that if she had won, that would have been important to her and to her well-being. When we lost I felt very disheartened because I felt we had a good case and I felt we actually had done a good job with the barristers and solicitors and experts. I think the court was worried that the flood gates would open. But, in the world of sexual abuse, which has subsequently become a more litigious area, clients are now coming forward some 20 or 30 years later and saying something happened to them when they were a child; they didn't realise what it was and they buried it in their subconscious, so they couldn't make a claim at the time or by the time they were 21, and the courts have displayed a more lenient attitude towards them.'

Glyn said this represented 'a change in fashion really, not a change in the law'. He added: 'So if Christine was to make her claim now, I think she would succeed. She did not succeed at the time, she was trying to break new ground.'

But Christine had legal aid and Glyn said this would be impossible now. It would take someone with money or sponsorship to fight this case now. But Glyn said he would like to try to take the action again. 'I would give it some consideration, but it would rely on very strong medical evidence and psychiatric evidence, which was prevalent in Christine's case. She had one of the top psychiatrists who was very supportive of her case and, as I said, her case did not win.'

I admitted I had sometimes felt bitter about the lack of compensation. 'I feel that all of you were not treated properly and if it was to happen now it would be completely different.'

I told him about the money given to bereaved parents, how it slowly rose from £250, to £500 and finally £5,000 whether you lost one, two or even three children. Glyn said: 'It is unbelievable, unbelievable, that those sort of figures were handed out to compensate people for the loss of children.'

Glyn said he was a young boy living in the next valley when the disaster happened. He said what always struck him was how people close to the disaster did not talk about it. I agreed: 'Parents never spoke.'

'Christine never had a conversation with her parents about that day, isn't that strange?' he said. 'And it's before counselling, it's before all that sort of stuff about how important it is to get it all out in the open. She said it was a non-subject, almost as if it didn't happen. No-one talked about it from day one onwards: there is the press of the world there talking about it, there is the royal family, there is Lord Snowdon, but locally people pretended it did not happen. Isn't that strange?'

*

I had known Christine Crocker for many years. She had lived for a long time in Perth-y-gleision, Aberfan, and I had bought Christine's family home after them. We lost contact when Chris moved away, although we had bumped into each other a couple of times over the years. We always chatted but the Aberfan disaster was never mentioned. I had followed the news articles about Christine's claim against the coal board but never thought we were getting anywhere near the full story. As Christine was not actually buried in the rubble, she was treated differently. She was given a terrible time for taking the action. Her treatment was unfair, and many people never really made the effort to understand her story. I really wanted to see Christine again and to talk about this huge issue, which had been missing from all our previous conversations. As Glyn has said, the disaster had been a non-subject for us. I was not quite prepared for how shocking Chris's story was, or for how much it would move me.

We met in the Courtyard Café in Crickhowell. Chris was

with her family, as her daughter was due to have a baby at any time. They thought that by taking her out it might 'spur the baby to come quicker'. When we met we hugged each other tight and sat with the sun beaming down into the courtyard. Then, in this genteel setting of home-made scones and Welsh cakes, Chris began to tell me her story. It shocked me, turned me cold.

As soon as we opened up about the disaster, we were able to immediately share our feelings about how much it still obsessed us.

'I am fanatical all the time,' I said. 'Something can't be right.'

Christine nodded. 'Our lives were ruined, Gaynor,' she said. 'They still are.'

I told her how I had spent the 30th anniversary in hospital. 'I really felt ill. Panic attacks. I thought I was having a heart attack.'

Christine said: 'The man who brought me through was a doctor named Choudrey. He is retired now but he was brilliant. He even came to my house for dinner. He cared.'

I told Chris that I didn't want her to go through her story if she did not want to. 'No, I am well now,' she said. 'I can deal with it. I want to tell you the full picture, the true facts, for others to know.'

*

Chris began to realise something was wrong when her daughter, Ruth, was about ten. Ruth was in school in Llangattock and Christine would walk through the woods near her school, sit on the grass and eat her lunch. She would do this every day. 'I didn't know why I was doing it,' she said. 'Anyway, this continued for weeks.' As a child in Aberfan Christine had often eaten her lunch outdoors at the senior school just before the disaster. Her actions in Llangattock were the signs of the start of a breakdown. 'Then one day Nick [Chris's husband] came home from work and he found me in the corner of the kitchen and I had torn all the nails on my hands out.'

Christine looked down at her hands. Her fingers tightened. 'I've got it now, look, but I can control it now.'

'Oh, bless. Love you,' I said.

'Nick phoned our doctor and he asked if there was anything in my life? And I said, "Yes, the Aberfan disaster". My husband even had never asked me about it, nobody talked about it. So anyway, he took me over to my mother's as he didn't know what to do. By the time I had got to my mother's I didn't know Nick, I didn't know anybody. I had gone back to being that child. The psychiatrist was called out and, in the meantime, I did a runner up to the cemetery. My father found me up there, but I can't remember it. They brought me down and wanted to admit me to hospital. My mother, who was a nurse, was there at the time. My mother took six months off work to look after me then. She took me to the hospital every day and cared for me at home. It got to the stage where I was fucking crackers, I was on so much medication. Ruth, my daughter, was ten; she is 35 now.'

Chris is 60 now. It was 25 years on from that breakdown but the feelings were still so raw. For Christine, realising that so many of her friends were under the black mass of waste, and scrabbling at it with her hands had totally transformed her.

'When I got to school it had just happened. I was on my way up to the Pantglas Senior School with the two friends.' Christine broke off, aware of her hands shaking. 'I jumped onto the tip. I was there all day, I didn't go home. I was digging, scrambling under the muck with my hands... deep under the muck. It was all the coal in my nails, so my brain is giving me the pain that I had there on that day... So all through my childhood I couldn't wash my hands. Look, you can see my hands now. I can't even do my nails now. I can't clean my nails out even today.' I can see the dirt under her nails. Her hands shake. 'So, anyway, things were getting worse and I had to be admitted. I don't remember it but I had three people come to see to me and they made me read all the newspapers for me to relive it. It still didn't do anything, I was that 11-year-old child. I didn't even know my children.'

Christine had to move out of the family home in Crickhowell as she tried to overcome her breakdown. She stayed with her parents in Aberfan for nine months, feeling she needed to be in the village to confront the past.

'I will tell you, it is incredible what the brain does,' she told me. 'The psychiatrist, he said he had been trying to get me out of this for three months and he couldn't. He said there was only one other option and that was abreaction. They don't do this lightly. They said it would only take one treatment and my parents and Nick discussed it and they thought if this was the only way they could bring me back, they agreed to try it. It was horrendous, Gaynor, fucking horrendous.'

I had to look up abreaction afterwards. Sometimes referred to as trauma therapy, it is a form of psychotherapy in which patients suffering from post-traumatic stress disorder are made to relive the experiment under the guidance of medical staff. This is meant to be a kind of catharsis.

'I had one treatment and everything started to pour out,' Christine said. 'What it does is it brings you down, with an injection, and just as you're going to sleep they stop it so that he could ask me, "Where are you?" And I was in the school in this sleep. Anyway, this went on and they asked if they could tape it as they have never seen this kind of reaction in anyone before. I had seven in total because each time I was under I would stop at a certain point. My heart was pounding so they had to bring me out of it. They taped it and asked my parents if they could use the evidence for teaching purposes in the psychological field of medicine. And now it is still being used in London, for teaching purposes.'

Christine put her hands out between us. 'I've got that feeling now, I want to tear my nails. I know what it is now, but my mind then was nuts. This abreaction drug was supposed to make you dream it in your sleep.

'Until you confront it, you can't move on, can you?' she said. 'My husband was brilliant, and my family.'

Christine's treatment took a total of seven years. She also told me that she had been to a psychiatrist as a child after the disaster. 'But they didn't pinpoint what was wrong with me.'

I said that my own experiences with a psychiatrist after the disaster had 'given me a complex that I was mad'.

I was only beginning to understand what Christine was trying to say. This breakdown had come as Ruth was reaching the age Christine had been during the disaster. Somewhere in her mind, she was thinking Ruth was going to be a victim. The abreaction revealed that she was, in fact, seeing herself as the victim.

Christine said: 'I eventually climbed a step and looked in. It wasn't Ruth, it was me! It was me, and my daughter was going to be fine. Mad, mad... Seven years of my adult life trying to get over it, but I haven't really.'

I was still struggling to feel Christine's pain. I said: 'It's never about me in my dream, it's never about me getting hurt or killed. It's my family.'

Chris replied: 'But it is you, Gaynor. If you went into abreaction you would see yourself. Your dream is you. You are seeing your life through the kids.'

'Well, that is strange because when I talk about the disaster and me on television or for the papers, I have always said it through a child's point of view or their eyes, as if it is not me, it is a child. I am talking as if it is another person, and that person is a child. But I can cope then, doing it like that.'

'That is what it is exactly, like me,' said Christine. 'If you don't deal with it like us, it can play tricks on you when you least expect.'

I said I was dealing with it by talking, by meeting people for this book. By seeking what I think is fact and closure. By finding the courage to ask my mam and dad questions I've never asked them. By getting into their minds' view of the day. My sisters too. And so many others.

Christine looks me straight in the eye. 'Why do you think they never spoke? Shall I tell you why, in all honesty? All our fathers were miners, and it was their livelihood. There was no help then if they didn't go to work. They were tied to the coal board.' Christine's face looks grim at the next memory. 'When I was getting well, I was horrible to my father, telling him I hated him because he was a miner.'

'Did you forgive him?'

'Yes, I did, because it wasn't his fault.'

*

So, how did the court case come about? I know Christine had a terrible time in the way she was portrayed for taking the action. I told her straight off that she did not need to justify anything to me about it. 'I am so proud that you tried,' I said. 'I don't worry about others. I wouldn't wish anyone the life we have had. I wouldn't wish what we have gone through on anyone. No-one can judge us, we all deal with it in our own different ways.'

She said she remembers the moment she managed to tell her husband that she wanted to sue the National Coal Board.

'I can see it now. Nick was sitting on the floor, I was sat on the sofa, and I didn't know how to say it to him. I don't want you to think I am a grabbing bitch, Gaynor, it took me days. I said to Nick, "I am going to sue British coal" and he looked at me and said: "What! I know what you have been through, but it is too late." But I wanted to talk, because Doctor Choudrey had given me the means of saying, FUCK YOU ALL! *I* am going to talk now.'

Christine went on: 'When I was getting well I wrote everything down and then I joined a disaster action group which I went to once a month with Nick. It was a cooperative thing, dealing with all disasters; there were Hillsborough victims there too. It was good to share our emotions and feelings, as they too had similar experiences. Then Glyn

Maddocks got a barrister and they were brilliant with me. And then it came down to the court case, which lasted a week. I wasn't nervous because I was holding it back. It wasn't about the money, I have to emphasise that.'

At the end of the case Chris said her legal team was very hopeful of success. She even had a call from another lawyer saying he anticipated that her case had opened the way for other victims to make a claim.

She waited three months for the verdict. She had lost. 'I lost because of those floodgates they thought that would open,' she said. 'But I wasn't sorry I had done it. I was trying to get my voice heard and stand up to the coal board.'

Christine felt there was still the possibility of a future action, but that it would need a group of people, a class action.

'It's time for everyone to stick together. We should all get together just like this, share our emotions, our feelings. I was fortunate to go through abreaction. I think all survivors should go through it, but it was hard. They should be prepared to give up years to come to terms with it; it will always rise up, it always rises up. It is never-ending. People tell you, "Don't hold on to it!" But we don't want to hold on to it; it holds on to us!'

Christine took me to her new home. A caravan near Crickhowell, on a site surrounded by trees and shrubs. There are mountains in the background here too, but they are green and safe.

I took some of the photos I had brought along out of my bag. We smile over a photo showing four of the surviving teachers, and we talk about people I have met. 'Some have faith, some have none,' I said. 'Faith in God,' she said. 'I lost all mine after the disaster. But when I was ill I lived in church, but I blamed God. I called Him everything, screaming in church for what He had done to me. But I believe again now. I have my faith back. I am better.'

Chapter Eighteen

A Town in Mourning

IN 1966, NO event as horrific as Aberfan had happened in Britain since the Second World War. The country had repaired since 1945. Veterans had coped alone. Moved on or broken down, silently. Aberfan created dozens of people who suffered trauma, were in pain. The country was ill-prepared to counsel us. There was little understanding of traumatic stress or of the needs of survivors and the bereaved. I'm told that even now there is no complete consensus on how to treat people caught up in such a disaster. How can any of us know what is the right thing to do?

Aberfan was a disaster unlike most others. The dead weren't from disparate communities brought together by the tragedy of being on the same doomed airplane, train or ferry; they were from the same small community. This was a huge tragedy concentrated into the area of a few close-knit streets, a couple of churches; these were the mothers, fathers, teachers, children, connected by one small village school.

In February 1968, a note-taker at a meeting at the Welsh Office took down the comments of an Aberfan social worker. 'The villagers had done admirably in rehabilitating themselves with very little help. A Government gesture was needed to restore confidence and only complete removal of the tips would do this. Many people in the village were on sedatives but they did not take them when it was raining because they were afraid to go to sleep. Children did not close their bedroom doors in case they should be trapped.'

In 1968, recording an open verdict on a bereaved mother

who died of an overdose of barbiturates, a coroner stated: 'I have no hesitation or doubt in saying that the Aberfan disaster contributed very materially to this woman's death. You can see the picture of her lying in bed ill, and the only solace and comfort to her in her illness was the photograph of the child she was clutching.'

Perhaps the adults suffered worse than the children in the coming years. For many there were drink, stress, disturbed sleep and psychological problems.

Aberfan GP, Dr Arthur Jones, said later: 'My work afterwards was more like that of a pastor. People had to face not only grief but bitterness and even guilt... It was predicted at the time that a lot of people might suffer from heart attacks brought on by stress and grief, but that didn't happen. Other experts predicted that there would be a number of suicides, but that didn't happen either. These people hadn't allowed for the resilience of the families involved. It was psychological problems that hit worst. From the time of the disaster for about the following six years, I dealt with people who suffered breakdowns. There was no set pattern or any time when it could be expected to happen. It happened at different times for different people.'

He added: 'After the disaster I warned that the community would have to come to accept its guilt. This guilt came out in many ways. There were the so-called guilty men who were blamed for what happened – they suffered themselves and were the victims of a hate campaign. But it wasn't only them. Women who had sent their children who hadn't wanted to go to school that day suffered terrible feelings of guilt... Grief and guilt came in many different ways. There was a strange bitterness between families who lost children and those who hadn't; people just couldn't help it.'

The misunderstandings in the management and attitude towards trauma, and the magnitude of the disaster in 1966, meant that the needs of the community, the survivors and the bereaved, went unmet. For many years, symptoms lingered and even today I believe we are still seeing the impact of the disaster in both physical and mental illnesses in Aberfan.

Back in 1966, psychiatric services were associated with asylums, severe mental illness, failure, weakness and signs of vulnerability. According to the modern theories of counselling, one of the primary needs of any disaster victim is to talk about what had happened. But for many of those in my community that simply was not what happened.

Our local doctors were put under tremendous strain following the disaster. One GP had lost a child himself. Two hundred and eighteen bereaved adults were living in Aberfan in the years after the disaster. Less than a third went or were referred to a psychiatrist. It is estimated that approximately 2,500 people took part in the rescue at Aberfan. Many took part in the recovery of bodies. Some were digging for the children of friends. Dozens were miners who would have to go back underground knowing that it was their industry which had caused this disaster. And there were teachers, traumatised, who had to go into the wreck of the school on the following Monday to recover text books and materials, such was the shortage in the local education authority. Who knows how many people altogether – rescuers, extended family, miners – should have got help for themselves?

In 1971, a follow-up prognosis of those who were suffering PTSD classified the recovery of 80 per cent of the adults as 'pretty poor', while 66 per cent of the children were rated good or excellent.

Further research in 1999, 33 years after the disaster, found that, of the 41 survivors questioned, a dozen continued to meet diagnostic tests for post-traumatic stress disorder. Twenty-five of the men and women, by then in their late 30s and early 40s, had experienced at least one symptom in the previous fortnight. Some had difficulty sleeping or bad dreams, and many tried not to think or talk about the disaster and to remove it from their memories. Psychiatrist Louise Morgan said: 'A few talked about the fear evoked at the sound of a lorry passing their house, or of an aircraft flying overhead. Intense memories are aroused by the slightest noise or smell. A number now have children the age they were. This seems to arouse new feelings.'

The agony I knew was felt inside the community and, in particular, the medical effects of that; it compelled me to speak to one of the GPs in Aberfan, to ask about their experiences in treating the people of the village. The GPs would have a unique insight from behind the closed door of the doctor's surgery.

Dr Arthur Jones had been our family doctor since 1951. Arthur was at the scene that terrible day; he understood the community, their feelings, the family trauma and the illnesses. Arthur always showed empathy to all his patients and he was a highly respected GP. He served Aberfan for almost 48 years. In 1983, he offered Dr Ramsewak Prasad a partnership in his practice in the village and both worked alongside each other until Arthur retired in 1993 and was replaced with Dr Shah. Sadly, Dr Jones has now passed away, but when I approached Dr Prasad he was glad to talk about what he had seen as the later consequences of the disaster on villagers, including my own family.

*

Dr Prasad had arrived in Britain in 1976 and had worked in hospitals and practices in London, Manchester, Newcastle upon Tyne and Durham. His medical specialisms included nursing, medical, psychiatry, geriatrics and COPD diseases. He recalls the offer to work in partnership with Dr Jones at Aberfan as overwhelming, as it was difficult to get into a general practice anywhere in those days and this gave him the opportunity to become a community doctor. Dr Prasad worked at the practice in Aberfan until 2008, so he had served for some 25 years as our general practitioner.

Dr Prasad was ex-directory, so I travelled up to his home, not thinking what I was going to say really. Trying to find his house was like walking through a maze. I was sent in one direction and then another by the local residents. It's funny how you can get so lost so close to home. Finally, I knocked on his door. I felt really calm by now, but there was no answer. I wrote a note on a scrap of paper, asking him to ring me so that I could tell him more about why I'd called. I posted it through the letter box but realised it was a rather cryptic, cold call and I didn't for one moment think he would respond.

A week went by before the phone rang and I could hear my mother chatting to someone and giggling. 'It's Dr Prasad,' she said, and I immediately felt a sense of relief. I spoke to him and arranged to meet him at his home the following week.

As I knocked on the door, I was feeling apprehensive. I am not a professional journalist or writer. For me, turning up at a house to ask someone if I could interview them for a book was a hard task. Scary, even. A fear of rejection has been forefront in my mind when I've asked all my interviewees for that initial favour. I am asking people to recall very personal things about a terrible event. This piles on my fear but I also believe it means that I have a good chance of success: there are genuine emotions at play here, ones which people want to share. There is a sense of wanting to extract some good from a bad event. As yet, no-one had said no, so I just hoped my luck would continue.

Dr Prasad opened the door and immediately shook my hand. 'Come in, dear, come in,' he said. 'Take a seat.'

I walked through the hallway into his lounge and began to relax. The delicious smell of his wife's cooking drifted in from the kitchen as he came to sit down with me.

I began to explain my story and asked if he would agree to me coming back to interview him about his experiences in Aberfan from a doctor's perspective. I think he was taken aback at first but he seemed very moved by my story and his professional relationship over many years with my family (and, in particular, my father), so with no hesitancy he agreed. I left him my earlier writings to read prior to our next meeting, to give him a greater view of the journey I went through as a child and teenager.

Preparing the questions I wanted to ask him was more complicated than the others as I have no medical training, so I just thought I would go with the flow. The taped interview took place in the lovely lounge of his home. I'm no David Frost and I was taping the interview using my mobile phone which I sat on the arm rest of his chair. I prayed it would work, as a few days earlier I had dropped it and the screen was cracked. I had brought two bags of sweets, mint humbugs and Opal Fruits, and I placed them on the coffee table. Dr Prasad told me he could not eat sweets as he had diabetes. Trust me to cock-up! I thought. 'Give them to your children,' I said and we both chuckled.

The conversation that followed was one of the most enjoyable I had while writing this book. It just flowed and I was engrossed by his responses; his view of the village was one I could get from no-one else. He was delighted to be involved. 'I practised in that valley for 25 years,' he said. 'It was imperative on my part to see you and I am very thankful that you would be considering me to take part. I am very inspired actually; I was very surprised to see you.'

He said he was moved by the 'attachment to your family, Aberfan itself and of course to Mr Minett, your father'. He paused. 'He was very, very close to me, you know? I remember giving him Zoladex injections [for prostate cancer] every month, very painful, in the tummy. He used to come to the surgery, always very optimistic and a very positive person.'

I smiled warmly at this assessment of my dad. 'He still is, Dr Prasad,' I said. 'He has had more than anyone's share of suffering.'

I did not know as I sat with Dad's doctor for many years that my father was in the last weeks of his life. His health was steadily declining as I made my journey through the past.

Dr Prasad had been informed about the disaster before he began working in Aberfan. He knew he was coming to a community where a generation was virtually wiped out. I asked him how he had been welcomed when he first arrived.

'What I felt was that people were very friendly, kind, generous,' he said. 'They accepted me. I didn't have any problems, nothing like racial discrimination or anything like that, you know. People were very, very friendly; they accepted me with a warm heart. So, I settled here quite nicely with no problems. I had no regrets coming here. My children grew up in this valley. My youngest was barely two years old, the eldest one was seven or eight. They grew up, had their education and studies in the Valley. I am still in the Valleys and I have retired now fully for seven years and I believe I will be here for some time yet, you know, until it's time to leave this planet.'

How did the work at Aberfan compare to his previous posts?

'The workload was tremendous because there were only two of us, myself and Dr Jones,' he said. 'At that time we noticed that the house-visit culture was very deep in the Valley. The patient would give you a call and, as most of the patients in the Valleys did not have cars, they would expect you to go to see them. So the two of us were actually on call at the practice 24 hours a day and seven days per week. You won't believe me, but on the weekends we used to get 30 to 35 house calls in a day for the whole weekend. We used to also do an open surgery on the Saturday up until 12 pm. There were no appointments, so we did not know how many patients were going to come. Arthur and I used to cope and made sure that one person was covering all the time, but it was a huge workload. Arthur, he was in his 60s then, was such a hard-working fellow, very sociable, very receptive, and so we had very pleasant moments. We never had any problems.'

Dr Jones had said that he believed at least 20 mothers and fathers died prematurely after the disaster and that he was still seeing the effects of that day in his surgery in 1991. I asked Dr Prasad what he thought of Dr Jones's assessment that even then, years after the disaster, Aberfan was seeing more medicines prescribed and higher incidences of illness than other comparative areas in south Wales.

'That's true, Gaynor,' he replied. 'The incidence of critical conditions were very high but, as it was a deep-mining area, there was a good possibility that people would have health conditions, especially critical conditions like lung disease and bronchitis of the lungs. You had this along with the effects of the disaster. Quite a large number of the patients, particularly the youngsters in middle-aged groups (they were around 30 years of age), used to suffer recurring conditions like reactive depression, depression and anxiety and anxiety neurosis, many psychopathological conditions, so the drugs prescribed in Aberfan were psychotropic drugs, sedatives, tranquilizers. Actually, the administering of them was very high at times. We had visitors from the local health board, some of the consultants at the hospital dealing with my patients there, and they used to impress upon us about the prescribing. They told us there should be some restrictions on us prescribing but the problem was that these patients were followed up at the psychiatric clinic at the hospital. They were seen at the clinic at the hospital and they were prescribed those psychotropic drugs by a hospital consultant and we used to simply abide by them. It was nearly always antidepressants. So quite a large number, or a good proportion of people in Aberfan, Merthyr Vale and Troedyrhiw, and the surrounding areas, were suffering from those psycho traumatic conditions.'

Aberfan happened before the accepted classification of the condition known as Post-traumatic Stress Disorder (PTSD), a definition which acknowledges that extremely traumatic events can produce chronic responses in 'normal' individuals; that people with no previous pre-existing psychological issues can suffer acute grief, anger, denial and detachment. Survivors of disasters and trauma had to rely on local support and whatever facilities could be made available by the local authority.

Dr Prasad said: 'I think the treatments prescribed at that

time were cognitive behaviour therapy and antidepressants. Those were the only two established regimen of pathological treatments that were suggested to my patients. And my patients were saying, "You know, doctor, they were not very effective". But you had only the two lines of treatment.'

He said despite the fact that little was known about trauma, there were warnings that the effects could 'go on for very long after'. He added: 'Some of the conditions were resolved very quickly, within a few months for some of them, but for others it lingered or continued for years longer, or even a lifetime. You can never put a time limit on this.'

I told him a little of my story. 'We still suffer now in our own way,' I said. 'Myself and other survivors still deal with it… but different people, different ways. My parents are still dealing with it and my sisters are still dealing with the impact in their own ways. The 50th anniversary is going to be bringing it all back, so I don't think the trauma, really, ever goes fully away. My mother was prescribed Valium and said she was in a state of being zombified, just out of it.'

Dr Prasad recognised that the medication prescribed to deal with people's health issues often did start a spiral of dependency.

'It was expected that if they went to the doctor, they would definitely be returning back with the prescription in their hand: that was their expectations of us as GPs,' he said. 'So then we had the next problem: to get these patients off those drugs because they became dependant on them and they would have withdrawal symptoms. And most of them, the patients, were also attending the hospital clinic and the consultants always kept them on [the medication]. As general practitioners we could not stop it. We were stuck with that, you know. And we noticed that the percentages of people suffering at the time with psychoneurosis, panic reactions and depression, what they call psychopathological conditions, were quite high in Aberfan itself. In the case of sleeping pills, we recommended it for a couple of weeks, but patients had previously been prescribed by the hospital doctors and were taking them for months and months, and years!'

I wondered if Dr Prasad, in the privacy of his surgery, had seen some of the things that I, as a survivor, had seen and experienced: the guilt the survivors felt. 'I remember we used to play away from the streets as we would get horrible looks from bereaved parents, which made us feel awful and guilty,' I said.

Dr Prasad shook his head gently. 'I did not notice any resentment or the feeling or expressions of resentment,' he said. 'It never came across to me. But the effects were still there and did not vanish in ten to 15 years after I came to Aberfan. There were people that never slept in the night.'

I said there was a problem in a mining community like Aberfan where the men did not want to seek help. They were miners! Strong, masculine men. They did not need help for trauma! At the same time the children learnt not to burden their parents or families with how they were feeling. 'The suffering can go unnoticed for many years,' I said. 'Many have died young, including survivors. There were those who drank themselves to death as they just couldn't cope with the trauma and suffering later in life.'

Dr Prasad sighed. 'I find it difficult to express my feelings on this, you know. I think that some youngsters at the time became drug dependant. There was a boy who used to come and I would not prescribe the drugs. Because of that the whole family moved out of my practice.'

*

Dr Prasad and I talked about the journey I was taking for this book. I told him that I was always trying to communicate with people because I felt there were those who could never understand what I and my family had gone through as they weren't there; and that I also felt that doing this would help me find an ending or closure. Failures by the Government meant that for many there could be no closure, as justice had not been seen to be done. Could we ever fully recover?

'I was not there, I cannot relate to the same experiences as you, Gaynor. You have 50 years of memories and feelings. You will go back and everything will flash in front of you, everything that happened to you. It is not a pleasant memory. It will open your wound. I think [the 50th anniversary] will also open up wounds for the people who live there. I will feel that too. I still remember well what Arthur used to tell me; how things happened and how he was involved in it. I think by putting your story in black-and-white you have done a wonderful thing; because it is history, and it is about resilience. It is very difficult to come to terms with, but I think this is your way of dealing with it.'

'When I look at others, who were not as positive and resilient as me, it has ruined their lives,' I said. 'Sharing my experiences, and my career, have been my saviours, as I have always worked in jobs for the welfare of others. It's my passion to help others. Giving back to communities something of what was taken away from me.'

Dr Prasad nodded, and then said: 'I think the effect of the trauma will be there and it will be forever-lasting and I don't think you will recover from it fully, you know.'

This was emotional for me. Because I looked to my parents to see how it still affected them in the latter years of their lives. Would it really be with me forever too? I said: 'I took my dad to the doctors to have his Zoladex injection yesterday and in the car he said, "Gaynor, why me, why me?" I don't get upset often but out of the blue that really choked me. I couldn't answer him, I was struck dumb. He has been a pillar of hope for the community and my family and has fought all his life to get the money back to Aberfan. Life is so unfair.'

'Gaynor,' said Dr Prasad. 'Your father is a true fighter, you know.'

Chapter Nineteen

The Pain is Just the Same

ANNIVERSARIES ARE PAINFUL for Aberfan. But they cannot be ignored. The disaster happened. Is the alternative to marking and commemorating, forgetting?

The 40th anniversary in 2006 was painful for my family and me. I will never forget that anniversary. There was so much attention on it and the build-up brought up the most agonising parts of the emotional journey over and over again. I remember walking alongside my dad up the steep hill to the cemetery for the 9.15am service. The weather was fine, for once. My sister, Michele, was walking on the other side of Dad. Suddenly, I stopped, I just couldn't walk any more: I was shaking so much. 'I just can't do it,' I said and I wept with emotion. To me, it felt like burying them all over again. The cemetery was packed with people, they had come from everywhere. I felt all eyes were looking at me. But I had to be strong for my dad.

Father Michael St Clair told the congregation: 'Yesterday I was asked several times, what do I think is the need for such a commemoration. We pray that all those who lost their lives are brought to that place of peace, light and happiness. Why do we commemorate so many things? Because we are a people who remember. Why are we here in these valleys? Coal is responsible. In those days they didn't realise what they were doing, in our days we are so concerned with global warming and the effect on the environment. What can we do to make things better for the future?'

Although it was intended that it would be a private service, the First Minister of Wales, Rhodri Morgan, was there. He said:

'Everyone can remember how they heard about Aberfan. I first heard about it on the lunch-time news that day and found it hard to make sense of the horror of it – the fact that a primary school had been buried, that so many children had died and the scale of the rescue operation. This was the 1960s. Wales still had 100,000 miners and coal was king. Coal was so important that we all accepted – until Aberfan – that there was a price to pay and we were all prepared to pay it. What we never foresaw was that coal could take the lives of our children. That was new, and terrible.'

The service made an impact on him, and I spoke to him to tell him how much my dad had fought. I was angry.

One of the readings at the service was by Jeff Edwards, a close friend and another pupil that survived the disaster. He was a councillor at the time and would later lead Merthyr Tydfil County Council. Thirty-four years after the disaster Jeff had experienced an amazing meeting with the man who saved his life that day. He was talking to another councillor, Roy Thomas, when they both realised that Roy – a former fireman – had carried Jeff out of the wreck of the school in 1966. Jeff had been the last child pulled out alive and had been saved when Roy saw his bright hair sticking through the dark coal waste. 'It was lovely to come face to face with the boy I had saved,' Roy said, after their reunion.

My dad was interviewed by the BBC on the day of the anniversary. He said: 'It doesn't matter if it's one year or 40 years on – the pain is just the same.'

*

My youngest sister, Sian, who was just a baby in 1966, gets very upset about any publicity surrounding Aberfan. The disaster robbed her of the chance to get to know her brother and sister. She can only listen to stories and look at the pictures that fill my parents' home. My mother had two portraits hanging on the wall of Carl and Marylyn. These are taken from their last school photo. The portraits are in oils and their eyes seem to follow you around.

When I sat with my parents to talk about that day, I wondered aloud if it had affected Sian. 'It was bound to,' said my dad. 'She won't show it, but that's worse.' Mam said: 'It has as Sian's got older, and she now teaches young children and has children of her own. She never had the chance to know her brother and sister, and can only look at photos and listen to stories and visit the cemetery, which was something she had not done until the last two years.'

Dad smiled gently. 'I can see Sian now, in the cot down number 17 Bryntaf. There in the corner. And Carl shaking the cot. I can see it now, as if it was by here.'

Another time Mam said: 'Remember the time, Cliff, when Carl took Sian out of the cot and took her out to the front pavement. I shouted: "Carl!" and he said: "I got her tight, Mam, I got her tight."'

The image brought a lump to our throats.

Mam said she remembered little Julie Regan. 'Julie used to love to come in and see Sian; she used to say: "There's a lovely dress, beautiful." Julie died in the disaster.'

*

Belinda is now 61. She was 12 at the time. She had passed her 11-plus exam. As there was only a pass or fail for the exam, if she had failed it she would have been still at the school, and would have very likely died as she would have been in the senior school where others perished or were seriously injured.

She remembers the 'chaos' of the morning in the house with all us children getting ready and baby Sian needing attention too. She does not talk much about the disaster either, only stating: 'Our family became closer, with our parents protecting us as much as possible from the grief and pain. Life moved on eventually, although our house was now quieter.'

Almost 30 years after the disaster my parents gave me Carl's football boots. The dirt was still on the bottom of them, from when he played his last football game for Pantglas School. It brought a lump to my throat when my daughter found these and put them on. 'Whose are these boots?' she asked. I explained to her, but she was too young to understand. Only in years to come could she fully realise.

My grandfather, Stan, who meant so much to me and who rescued me from Pantglas, sadly developed Parkinson's disease. There were so many things I wanted to ask him but never plucked up the courage to. He passed away suddenly, with no time to tell him that I loved him. I was so sorry afterwards. Why do we always make the mistake of not telling our loved ones how much they really mean to us? It's always too late afterwards.

At least I did not make that mistake with my father. My dad never spoke to anyone about his feelings, but time had healed some of the scars and only when I started my journey did he feel able to talk. My parents always visit the cemetery and my father was on the cemetery committee for many years.

I was so glad that I got to speak to him about his life when I did.

Chapter Twenty

Goodbye, Dad

OVER THE COURSE of two years Dad's health deteriorated. Once, he lay in a coma due to sepsis and the doctor warned us that the next 24 hours were crucial. All seemed so unfair; he didn't deserve that. Reality struck me then: I may never know the truth of what happened to him. I had to ask those unanswered questions. I wasn't being selfish: for me, he was the key to my journey and only he could unlock the answers.

Within five days, Dad had pulled through. It wasn't his turn then; he was our fighter, just like he had been for almost 50 years. A true hero. But the prognosis for him was bleak.

As the following two years went by, Dad became less mobile and much weaker, but he remained positive. His daily job was feeding the birds. He just loved them. On many days he would sit there with his binoculars, watching to see which species they were. He would spend lots of money on the bird feed, they had the best. He always said they were the best fed birds in Aberfan.

Us girls bought him a bird house. It was placed right outside his conservatory. Yellow flowers surrounded it. Dad sat in his armchair, which my grandfather had sat in before him, and would watch the birds all day.

I dreaded the day my parents would leave this world. Memories are golden and my memories flood through my mind all the time. I will never forget the last few weeks of his life. We shared laughs, cries, emotions, and a most special moment just four days before he died. Dad had spent a week in hospital before being sent home. We all cared for him in the conservatory, where a special hospital bed had been placed, so he could watch his birds. All Dad's family came to visit, his sisters, brothers, friends, and all his grandchildren and great-grandchildren.

This particular afternoon Dad and I reminisced, just us. I sat by him, stroking his head. I had lived with my parents for roughly a year, he loved having me there. I was just due to go back home. But this moment was special, and he asked me to stay longer. 'You will be safe here, Gaynor,' he said. He had known about the book and the people in it, but then he said, 'Oh, Gaynor, and your book.' I promised him I would stay.

Those who truly know me and knew my dad know pretty much every heartbreaking detail of the pain I have endured after losing him. It was the worst day of my life, but in a strange way I was so happy that he would eventually be at peace.

The dreaded day approached. Heartbreaking. All the family were around him in his final hours, in the conservatory, by his birds.

I grabbed his hand, stroked and touched his hair. He was sleeping, but I knew he could hear us. We even sang to him, his songs. 'If I were a rich man...' There was a huge sense of helplessness. Some things will forever be out of your control. There was nothing we could do now but let him know how loved he was and will be forever. As Dad took his last breaths, we all said we loved him. He was showered with love. Mam held his hand and said: 'Go Cliff, go to Carl and Marylyn, give them a big hug.' I will never ever forget those words she spoke.

*

My father died on 25 July 2015. He and my mother had been together 61 years.

Those days after, before he was laid to rest, seemed like years. We lived next to the cemetery and Dad had always said 'he would just go over the wall', but this time it was real.

Dad's bedroom window had faced the cemetery. He and Mam could see the arches where the children were laid to rest.

For many years, I had always dreaded the thought of Dad or Mam being buried, the thought of the graves being opened. Dad was to be buried with Carl, Mam with Marylyn. I will be with Dad and Carl when it comes to my turn.

We had planned for Dad to come home to the conservatory ahead of the funeral but we had to make last-minute changes. Dad would always give us problems, so we were used to this and we thought it funny, really. Family, friends, doctors, politicians, they all came. The weather was fine. Dad was in the hearse, right by his bungalow, just outside the steps that he had only cemented a few months ago. The family were all around him in the garden, the weather was fine. We played Dad's favourite music. The lay preacher spoke about Dad, all about Dad. As Dad wasn't religious, it was about him, the fighter, the hero.

We then took Dad to the memorial garden for a few moments to pay our respects. The undertaker walked in front of the hearse. What an honour.

Dad had spent his life serving the community: he worked in the gardens, cemetery, making sure all the work was done. He spent many hours working there. This was the place where his story all started. The memorial gardens, where Pantglas School once stood.

We said goodbye to him at the cemetery on the hillside in Aberfan; among the graves of the children. With Marylyn and Carl. It would have been his 83rd birthday.

Arriving at the cemetery, we became almost blinded. You don't take it in, can't believe it is happening. There were people everywhere, but I don't remember who they were.

Dad loved his family and considered himself a lucky man to have eight lovely grandchildren and seven great-grandchildren. All Dad's grandsons carried him that day. They were strong, proud. Ross, one of the grandsons, read the eulogy. His voice told the story of what an amazing man he was. He was loved every day of his life and he knew it. There was some humour, too. We smiled; we had to.

Dad was laid to rest, and I was in control until that point, then I gave an almighty deep breath and suddenly felt like I couldn't breathe. I felt like time had stopped. I shouted out, 'Oh my God', and tried to control my breathing. I was helpless. Suddenly Dad was gone.

*

A few weeks after the funeral I asked Mam if she would write a few words for my book. Mam went outside and sat on Dad's favourite bench. As the tears flowed she wrote, 'I am Iris, mother to Belinda, Marylyn, Gaynor, Carl, Michele and Sian. I have just come to terms with burying their dad with Carl by Marylyn's side. A happy house that morning, only half a day. I said to them: "We will have Halloween fun, no school for a week." Little did Cliff and I know. Coming home that night after seeing Gaynor with her leg broken and asking for them... Oh, I don't know how my whole body didn't crumble. Cliff, Cliff, my soul mate, we struggled to carry on, our love for our girls and each other was the bond that kept us going. Flowers, flowers, damn flowers, that's all I can give you three now.'

My dad loved my mother to bits. There was a lot of hugging and kissing in the home we all grew up in. My sisters and I were given the building blocks to build our own lives for which we were truly thankful.

There have been many tears, hugs, words of comfort within the home since Dad died. Us girls have spent much time chatting, remembering, shedding tears.

Belinda said: 'My father battled with Government bureaucracy on behalf of all the bereaved whilst trying to support our family. Over the years my parents supported, cherished and loved each other unreservedly and for this we are all forever grateful. My father will always be remembered by us and many others as a determined, passionate man who cared about the people of Aberfan.'

Sian said: 'I was born seven months before the disaster, the youngest of six which became four. I am 50 in 2016. My childhood memories are filled with love from my parents and sisters. Just like all the other families, we lived with grief and loss every day. Yet I was lucky to have strong parents who spent all their time with us. As we age we realise the gift of time is precious – my mam and dad gave this to me. After each and every working day, I was at my father's side, I went everywhere with him, and I was at his side when he passed. Who can ask for more?'

Michele spoke to me about how she cherished her friendship with her sisters, and how she had been guided by the love of our parents. 'It is only when you lose someone so close that you realise that the love and happiness our family has will always remain. Mam and Dad were our rock. They gave and inspired us with their unconditional love, support and guidance along any road we took. We knew they were there. So I say to Dad: "Love you, miss you. We're coping one day at a time."'

We are lost without Dad, but his memory will live on forever in our hearts and minds. He is not here to read this story of my journey and, for me, that is so sad, particularly as he was such a part of it. But he will know as he looks down from heaven; he is still around us.

Love is stronger than death.

I knew that one day Dad would make this journey but still, to be honest, I had not been ready to say, 'Goodbye for now'. Now I must...

Goodbye Dad, till the day we meet again, when we are united. Love from us all.

Chapter Twenty-One

A Question of Faith

DURING MY FATHER'S long illness time just seemed to stand still. We were waiting for the end, I suppose, while at the same time not wanting it to come. During those months, June Vaughan was one of the people in the village that I truly looked forward to speaking to. She has long been one of the spiritual leaders of Aberfan; she is humble and compassionate.

I had always pondered about faith and how it had influenced me in my healing. It had seemed to take me out of my negativity when I had found myself in times of doubts and uncertainties. I wanted to speak to June because our conversations brought with them an almost inexpressible comfort of feeling safe; she was someone I could trust when speaking about such sensitive feelings. Her positive contributions to the community then, and still today, have undoubtedly had an impact on those who came into contact with her, myself included.

I took a bunch of flowers to June, white carnations as a token of appreciation; I wanted June to know that she was special in my heart.

I arrived at her home and, before I had walked through the passage, her first words to me were: 'I am so sorry for your father, Gaynor.'

As we sat in her homely living room, June – now 84 – said she had many memories of Dad, including the very poignant memories of him attending the memorial services at the cemetery and of the practical work he carried out at the memorial garden. My own abiding memory of June is as this petite 'lady reverend' who, for many years, led the memorial services at Aberfan cemetery on 21 October every year.

June said: 'We have the service in Aberfan cemetery at 9.15am every year on 21 October, which is the mayor's service attended by Merthyr council, police, residents, dignitaries, bereaved and anyone wishing to attend. Then, in the evening at 7pm, we alternate the memorial services between the four churches, Merthyr Vale Baptist, St Benedict's Roman Catholic, St Mary's Church in Wales and Zion Methodist. We have been doing this for a very long time, as far as I can remember, and I usually do what they call the commemoration, because I am one of the few that is left from the time. Sheila Lewis has read from the Bible in that church service on many numerous occasions. We are going to still continue to have that service for as long as we can. About 70 come to the service. What we found is if people hadn't come (speaking mainly about the bereaved now), if people haven't come years ago, they are probably reluctant... Or I put it this way: some people find it easier to bear at home than to come out to a service on that day. I can understand that people may not want to come out on that evening, I am able to understand how the bereaved feel. I didn't lose a child and it is not the same, you see.'

I said my mother had never been able to attend a service and that she stayed at home. 'To this day she cannot even go to any children's concerts. Once she tried to, and she had to come away, it was too unbearable for her.'

June said that some parents would attend one service but would be unable to come again. 'Somebody once said to us, "Why are you keeping on with the service, don't you think you should give it up?" We said that while half a dozen wanted to come, we would still do it.'

'For years I did not attend any service, it was too much to for me,' I said. 'But when I didn't go, I would feel very guilty so that made my feel bad too. So I decided to attend. Some years are more painful than others. Some years I attend and don't cry, others I just cry uncontrollably. I am now dreading this year's service as Dad will not be there.' For many years Dad had laid a wreath and he always made sure the wreaths were laid in the memorial gardens and the cemetery.

*

June Vaughan was 34 when the Aberfan disaster happened. She worked as a local preacher and later became the minister at Bethania Chapel. June met her husband-to-be, Cyril, when she was 14 and lived in Mackintosh Street, Aberfan. June married at 21 and moved to Birmingham where Cyril was teaching. Soon after, Cyril's mother had a stroke and they returned home to look after her. Cyril had a teaching post in Ebbw Vale and then a job became free at Pantglas Junior School. After ten years at the junior school Cyril then went to work at the senior school next door.

In 1966, June was a member of the Red Cross, and she looked after her three children: Michael, who was six, and in Perth-y-gleision Infants' School, Peter, then four; and her daughter, Judith, who was two. Peter is now Chief Constable of South Wales Police.

June told me that on the morning of the disaster, Cyril took her to Merthyr Tydfil for dental treatment. 'While we were there we met a friend, Bill Evans. We chatted briefly, and Cyril said he was going to be late for school. We were later told that Bill, who lived next to the school, had lost his wife, two children and his home.'

June went on, her voice filled with emotion: 'We arrived back at Aberfan at 9.20am and saw distressed children on the main street. "The wall has fallen down on the school." Cyril went to the school. I went to my parents' house where I had left the two younger children. I walked to the school and could not believe nor understand what had happened. Most of the school had collapsed. I could see a mound of black slurry, black coal waste. To my horror I realised that the tip had slipped.

I climbed into the school through a window and along with others moved rubble to gain access to the classrooms. Then the police, ambulance and civil defence were there, followed by the miners from the colliery. There was a great sense of relief. They were trained in rescue work, they would know what to do. Later that evening we had a telephone call from my sister and brother-in-law in America. We asked ourselves how they knew so quickly, unaware that, for the most devastating reason, the name of the village of Aberfan had travelled worldwide in just a few hours.'

The following day, it rained incessantly. June worked to give food to the rescuers. She remembered one 16-year-old boy who had come from the Midlands to help; the boy worked continuously on the site. 'We served soup from the front gardens of the Moy Road houses,' she said. 'This boy said he didn't like lentil nor tomato soup. I assured him it was neither of those, it was a mixture of all kinds. "Best soup ever," he said. I will never forget that boy.'

For June, it was her faith that helped her through that time. 'I need to go to church, it's all-important to me. At one time we had 11 churches in Aberfan and Merthyr Vale, now we have four, Methodist, Church in Wales, Roman Catholic and Baptist, with few attending.'

June was able to provide support as she knew all the families in Aberfan but she was in a difficult situation, gauging how to preach Christianity to the bereaved. She focused on practical support and words of comfort. I told her how many a parent's religion was tested, how some became atheist, their faith changed. It was not just the disaster, it was the cruelty of it. Half-term was just one day later. None of the children would have been in the school. June said, 'I don't know how the parents and grandparents managed to keep going, such was the enormity of the situation. I really don't know how they faced up to it. The courage and dignity, and then the time of the tribunal, must have been overwhelming, the heated discussions, disagreements, and yet they faced it with that same courage and dignity. I have asked friends how this was possible. They said they helped one another, that they were not alone in their loss.'

One bereaved father told a television documentary in 1996: 'As far as we're concerned now, we've still got two boys. We're only separated for a time. One day we're going to meet. The parting and the loneliness and being without him is terrible, but it's not for ever.'

June said that she spoke to a bereaved mother who is now in her 80s about her daughter who died. The mother had said: 'I keep reminding myself to relive the love, the happiness, the fun of the years we shared, and that makes life easier for me.'

June added: 'She believed she would be reunited with her daughter one day. This has given her hope and comfort throughout the years. It is our Christian belief that this life is not all, it's just a stepping stone to a richer, fuller life where we will be reunited with our loved ones in a glorious reunion in the care and protection of God our heavenly Father.'

*

Perhaps because June knew so many people from the community, we began to discuss the people I had met on my journey. 'Sheila Lewis's contribution is so important because of the loss of her child, then the loss of her son. So traumatic,' said June. 'Sheila is just great. Please remember me to Christine Crocker. I knew her family.'

I told her about Francine Jones. 'Oh, yes, I remember. You are telling me so much news about these people. I have always wondered where they are, what they are doing. Wonderful to hear about them. I now know they are all OK.'

'It's been quite a journey, June, to find out all these things,' I said.

We talk about Francine's story. June said: 'Can you imagine what it would be like to lose your mother, father, grandfather?

What on earth did that child go through at eight years of age? And not to die in hospital in a comfortable situation but to die in their home!'

I shook my head slowly. 'My heart goes out to her. My life doesn't compare,' I said.

I asked about the Aberfan Wives' Group, formed in the wake of the disaster for the women of the village. The men formed a choir, the women the wives' group. Both continue to this day.

A few years ago the wives were invited to a Buckingham Palace garden party. 'There was a group of about 24 of us and I was to be one of those presented to the Queen,' June said. 'The Queen came out first onto the balcony and they played "God Save The Queen". She looked really beautiful, really beautiful. And I said, deep breath, "Oh my gosh, the Queen is here. She is in all her splendour and pomp". She came down and I think Pat Lee was one of the first to be introduced. The Queen said to Pat, "You lost a child", and Pat said she had lost her daughter. And the Queen spoke to Pat and she changed from being the Queen of the country to being a mother. Then she asked me how the wives' group was formed and I said it was as a direct result of the disaster.'

June added: 'There is no way that you can keep the 50th anniversary low key. You have to be realistic about it and it's going to be something very, very big... The disaster had publicity worldwide. Because it was children, who should have been safe in school. If it had been the miners in the colliery it would have been horrific, yes, but because it was children it is so sensitive. I was very apprehensive about talking to you, it's only because I knew you so well that I did. I felt apprehensive because I thought, because I did not lose a child, am I entitled to speak about my views? When you said to me will I do it, my thought was no, I was not in the position to do so, it should be all the bereaved. But I didn't realise that you were doing the whole spectrum of it.'

I told her that, for me, talking to her was very relevant

because Christianity had helped get me through my journey. I said this story was a large and intricate jigsaw for me. I had to have all the pieces.

Seeing June reminded me of a minister called Mr Penberthy, who ran the village Sunday school and a Monday evening club known as Sunshine Corner. I remember going to Sunshine Corner, singing and drawing and having so much fun. More than 20 children from the group and 14 from the Sunday school died in the disaster.

June smiled: 'It was Mr Penberthy who asked me had I thought about forming a wives' group. From there we went from strength to strength and I was the first president of the wives' group. If there was ever a man that was heaven sent to Aberfan, it was him. His concern and interest for the young generation of Aberfan was remarkable.'

As the time came to leave June, she walked me to the door and said: 'Come here, Gaynor, let's have a hug. It is good to hug. We don't do it enough.'

We embraced each other. The feeling was good. Unifying, sealing a bond even stronger. I felt so fulfilled speaking to June. We shared a common faith. Her sincerity just made you feel that life was good and you had to live it to the full. 'Bye for now, June,' I said, and waved to her as she stood in her doorway.

*

I picked up the phone and dialled the number, not knowing what I would say. In my mind I could vividly see Mr Penberthy as he was all those years ago, telling us about 'Billy Blackbird' – a gentle story he made up to comfort us after the disaster – or driving the little minibus which would pick up all the children from Aberfan and take them to Sunshine Corner and the cinema club.

A quietly spoken voice answered. 'Is this Mr Penberthy?' I asked. 'I am Gaynor Madgwick from Aberfan. I used to be Gaynor Minett from the bungalow. Do you remember?'

'Yes, yes, I remember. Some of the names have changed like yours, "Madgwick".'

'I got divorced quite young but I kept the name Madgwick,' I explained. 'It is so lovely to speak to you after all these years.'

'It's lovely to hear your Welsh accent,' he said.

I explained that June Vaughan had given me his number and why I was ringing. He told me he was 84. 'I hope I continue to be preserved so I can come to the 50th anniversary,' he said with a deep sigh. 'I came to the 40th and, at that time, I thought I probably wouldn't be around to come to the 50th, but I am praying!'

We laughed together. 'Your praying is doing you good,' I told him.

'Well, I so hope. I pray for you all there in Aberfan. I haven't forgotten you. I will never forget you.'

His voice began to break and I felt a lump in my throat, too.

'I am sure you have had a job tracking everyone down,' he said.

'Yes, I've been privileged to speak to everyone.'

We shared a laugh about our memories of Sunshine Corner.

'And then there was Saturday morning cinema,' he said. 'I bet you remember I played the piano accordion there. I still play you know. Some of the music that I had written, we performed it after the disaster, and I recorded it. I still have the recordings of some of the Sunshine Corner children singing. And, of course, I used to go down to Church Village hospital to see the children and I also came to St Tydfil's hospital to see the children there. There were a few at St Tydfil's and four, I think, in Church Village.'

'I was in St Tydfil's for three months,' I told him. 'Francine, you would have seen her at Church Village. I met with her a couple of weeks ago and she is doing very well, she is happy.'

'I have her singing on tape,' he said. 'I have Francine singing in Welsh on the CD I had made. I will be glad to send it to you.'

I told him that June Vaughan had described him as someone who had been sent from heaven for Aberfan.

'This was because I had this contact with the children through Sunshine Corner, but then, of course, the next summer...' He went quiet.

I realised I had opened up Mr Penberthy's memory with my unexpected telephone call. I hoped he felt OK on the other end of the line.

I had a newspaper article in my cuttings file which reported that he had gathered his congregation in the Methodist chapel on the day after the disaster. He had told the reporter: 'My own faith has been shaken to the core, and I like many others, needed a few quiet moments to think about what had happened and to try to restore our faith in God.' There were no hymns that night. 'The first verse would have stuck in our throats.'

The reporter noted that Mr Penberthy's attention was taken by one empty corner of his chapel. 'They would have been here at Sunday school this afternoon, over there, two little brothers sat together,' he said. 'They died together.'

I told him June had said that he made things much better for the children after the disaster.

'I prayed a lot and asked the Lord, "Help me!",' he said. He stopped that memory there, and said: 'Oh, Sunshine Corner. We used to have great times!'

Mr Penberthy remembered the song we used to sing:

Sunshine Corner, oh, it's jolly fine,
It's for children under 99,
All are welcome, seats are given free,
Aberfan Sunshine Corner is the place for me.

That took me right back and I couldn't stop laughing.

'I remember the minibus picking me up from outside my home,' I said.

He laughed. 'Oh, yes, I had a loudspeaker on top of the minibus at one time announcing the Saturday morning cinema!'

I had never known his first name in the old days. He was always Mr Penberthy. 'It's Irving,' he told me. 'When I was learning to drive they called me "swerving Irving"!'

We laughed again and I told him: 'It's so lovely to hear your voice.'

'May you be strengthened and comforted,' he said. 'I have sad memories, of course, but I have many happy memories too of living in Merthyr Tydfil.'

I told him about how faith had 'healed' me, but that my parents had become atheists.

'It's a pity we blamed God,' he said. 'We should have blamed the coal board.'

My mother said God would have parted the sea of slurry, I said.

Mr Penberthy sighed. 'I can understand them but even Jesus died on the cross: God didn't stop it. This is a troubled world and there wouldn't be any need of heaven if everything was all right down here. Down here we are being trained; we are apprentices learning the ropes and it is very painful sometimes. I bet you have learned a lot too from your ups and downs. How old was your dad when he passed?'

'Eighty-two.'

'You know what the Bible says? Our span is three score years and ten. You can look back and know you have looked after him, done your bit. God bless you.'

As we say goodbye he says one final thing. 'Bore da, Aberfan.' Hello, Aberfan.

Chapter Twenty-Two

Pandora's Box

THERE IS A quote by the young British-Indian actress Katrina Kaif which says: 'Going by my past journey, I am not certain where life will take me, what turns and twists will happen: Nobody knows where they will end up. As life changes direction, I'll flow with it.'

I have found from the years of struggle that we do not take a trip: a trip takes us. This was not a journey of discovery for me, it was a journey of recovery. My solo journey started out within the village of Aberfan with no set itinerary. I had been taken to meet some amazing people, visit places I would have never imagined going to, and listened to stories that have been beyond my imagination.

The questions I had asked of myself and others over the months had given plenty of answers, but I found that they did not quench my curiosity to discover even more. I was given retired policeman Alan Roberts's telephone number after he had a chance meeting with a friend of mine. He had stumbled onto my path and would now become part of my journey.

As I planned for meetings, I became aware that I would not only be listening to new stories which would aid my understanding of that day in Aberfan, but that the conversation would also be of help to my interviewees: I was meeting people who had a past to unlock and unload. They had carried these stories with them for 50 years, never having the opportunity to share them: it made complete sense to finally share them with someone who was both a survivor and bereaved. Reflecting on this increased my compassion and appreciation for all those who had shared their experiences with me.

Meeting Alan took me to a place with many happy memories, Pembrokeshire. The little coastal village of Saundersfoot had been a place of escape for many years. It is peaceful and breathtakingly beautiful. Whenever I can, I escape to my caravan there. Alan lived only a few miles away in the equally-quaint village of Penally. This too was a place I knew well. When I was a teenager, this was where we always came for our family holiday. As I drove to Alan's house, I even passed the house where Dad, Mam, my sisters and me used to stay, and the Cross Inn where Dad enjoyed a beer or two. It brought a lump to my throat. My dad had died only three weeks earlier.

Alan's cottage overlooks Tenby Golf Course and the sand dunes which run from Tenby's South Beach to Giltar Point in Penally. My sisters and I used to love to play in the dunes. It was strange that fate had provided two locations which Alan and I shared, and not surprisingly we connected immediately. The two of us were joined by his wife Betty and we sat at a dining table, looking out over his garden and that view to the dunes beyond. The flowers were in full bloom; it had been a summer of changeable weather but the sun was shining brightly now. Alan, though retired, still had the physique of a policeman. His wife, Betty, is dark-skinned and lovely. I thought she had a Spanish skin tone. 'No,' she replied. 'But everyone asks me that.' Betty smiled: 'Do you want a cup of tea? I only have goat's milk.'

I had never tried it but it was the best cup of tea I had ever tasted.

Alan knew nothing about me, so I explained my story, the range of people I had met, the fact that the youngest had been ten and the oldest in her 90s. I didn't want Alan to feel intense or concerned that anything he might remember might upset me, so I told the light-hearted story of how I almost missed my appointment with Lord Snowdon. Alan and Betty both laughed. 'Your story sounds incredible,' they said.

I replied: 'But I wouldn't wish it on anyone.'

We talked about the silence that had surrounded survivors in the years after the disaster. 'It tended to be like that in those days,' Alan said. 'People didn't speak.'

Betty said: 'Do you know my mum died when I was 14 and my brother and my sister never talked about it. I find that very difficult, because the thing is right there in front of you...'

Maybe there is something about grief, I think to myself. Something about the fact that we dare not speak it out loud for fear of breaking down. Or is there something in us that tells us that if we don't speak about a catastrophic event then maybe we can persuade a part of ourselves that it never happened?

In responding to his wife, Alan also seemed to pick up on my train of thought. 'Yes, it's like having your own Pandora's Box. As a policeman you become involved in many things, happy, sad, good and bad. If you were to tip out what's in my head, you wouldn't be able to put everything back in... It's that sort of thing, isn't it?'

I told them about how my dad had spoken about the book I was writing in the final days of his life. It was his regret of not being around to see it finished.

'So I thought I have to finish it now.'

'And you will, won't you?' said Alan. 'Because you have too.'

I told them I was using one of my father's favourite sayings for inspiration on my journey. When I had been talking to my parents about the book and their lives since the disaster my father had given my mum a *cwtch* – a lovely, warm hug – on her birthday when he said: 'Oh, Iris, we've done it one day at a time.'

One day at a time. Over half-a-century. That's how they got through, recovered, from Aberfan.

Betty said: 'So you are being pushed now to finish it. He is that drive inside you.'

I said it was hard to lay myself open in a book like this, unsure of the reaction, particularly in my own community. 'It is history, an important part of our social history,' Alan said. 'So many good things will come out of it. What you are doing is essential, the story demands to be told, especially by someone who was there. And never forgotten.'

Then Alan told me his story.

He was living in Bridgend at the time of the disaster and working as a scenes of crime officer. He was 23 and had been in the police for four years. He was based with Glamorgan Police, one of a number of forces in south Wales at the time.

'You will know that Merthyr had its own police force when it happened,' he said. 'And of course the event was of such a catastrophic nature that everybody reacted to what happened, with many phoning to ask, "What can we do? What do you need? I will help", and Merthyr Police switchboard just couldn't handle the number of telephone calls received that first day.'

Police officers in the neighbouring forces took many of the calls, including Alan and his colleagues in Bridgend. 'We would take details of what they were offering and how we could get back in touch with them if needed. That was on the first day, the day that it happened. It is difficult to remember any individual call, as there were so many and from all over the world. You know, people were ringing and saying, "I know I am miles away". Some were many hundreds of miles away, but all of them they asked if they could be of any assistance…'

That night Alan was told he would be going to Aberfan the following morning. Six officers made the journey in a police van. Alan remembered driving into the town. The roads were 'almost impossible' to drive through, but they got to their destination: the temporary mortuary at Aberfan Chapel. Just weeks before Alan and I met, the chapel had burnt down. Only a shell was now left.

Alan explained that some of the smaller police forces had insufficient resources or manpower to deal with the complexities of such disasters back in 1966. 'Merthyr was one of the smallest forces in Great Britain and that caused its own difficulties in that there were very limited resources that they had to call upon, so it was necessary for them to request mutual aid from other forces, especially those close to Merthyr,' he said, 'and we came to assist in whichever way we could.'

I said communication must have been difficult full stop, with narrow valley roads busy with rescuers and people wanting to help, and with no mobile phones to help coordinate the operation.

'In 1966, communications could be a problem but land-line telephones could be temporarily installed and that helped,' Alan explained. 'So, for major incidents like this, wherever they were needed, temporary lines were installed within a short space of time.'

Alan worked for just under a week in the mortuary. He said people from the Salvation Army and the Red Cross were doing 'outstanding' work there.

Did he think it affected him?

Alan took a deep breath. 'Over the long term, I don't think it did, but you couldn't experience something like that without it having an effect at the time, but you couldn't dwell on things. I had a job to do and so I got on with what I'd been asked to do.'

It is not easy for Alan to forget his work in the mortuary. 'That Saturday, I came away from there having seen what I saw in there... and particularly the folk that came in to try and identify their children, or whoever was there, and to see those people absolutely devastated... When I went away that night, I must be honest, I bawled my eyes out, but that was good for me because it was a release and it helped me. And so the next day, as a policeman, you have to go back in there. You can't say, "I can't do it".'

Betty said that Alan was able to combine being a policeman and a family man; that when he put on his uniform, his back straightened and he was a different person. 'But I will say, he does not talk about Aberfan. Ever. It does affect him.'

This is a conversation, I suppose, that Alan and Betty had never had. It was part of Alan's Pandora's Box.

'I know I have never spoke about it,' he admitted. 'Other than at the time when I said that I bawled my eyes out on that Saturday night.'

Betty said there may have been a greater shock to it at the time. 'You didn't see things like that then; now you see it on TV. People are getting more adjusted to what they see. The people who saw this... This is real life...'

'Also many of them were miners,' said Alan. 'And whilst they might have not experienced tragedy in their lives before, they were mentally always geared up for it because tragedy could happen at any time with that sort of job, where they never knew if they were going to go home.'

While others were around, he said, they had a job to do and they did it. If there were to be tears, like him, they would have done it somewhere quiet, somewhere alone. 'You have to do that away from other people and I am sure all of them must have done exactly the same. But when they came into that chapel and had to face whatever it was that they had to face, they were like the policeman and fireman, they were the brave, proud miner, and just as I did, they must have said to themselves: "I've got to deal with this and I've got to deal with it properly." Yes.'

We took a breath as Alan had to make a phone call, and Betty talked to me again about my 'drive' to finish the book. 'I could never stop this book now, I would be ill if I did,' I replied.

When Alan returned I asked if he had ever been back to Aberfan. He said no. Would he go back for the 50th anniversary? 'No, I don't think so. My involvement was fairly limited; it was a very precise involvement. As much as it affected me, I knew I had to move on from that. There was no counselling then, you just got on with it and you never talked to anybody about it. No-one would come up to you and say, "How you are feeling now?" They just said, "You are needed back there tomorrow, for 7am", and that was it.'

Chapter Twenty-Three

More than Just a Name

THE STREETS WHERE we all live in Aberfan have changed little since the day of the disaster, but the geography around is different. For a start, the main Cardiff to Merthyr road now runs on a dual carriageway on a hillside above the village. This road, the A470, cuts right through the path of the disaster. The sliding tip would have crossed it had it been there 50 years ago; now cars speed along it, with Aberfan down out of sight in the valley. The passer-by would have no idea how close they are travelling to the scene of a disaster.

Merthyr Vale colliery closed in 1990 and its site has been landscaped.

The tips have gone and where once there was darkness, there are pastures green with trees that have grown once more.

On leaving school I gained five CSE examinations and then went to work in a solicitor's office in Merthyr Tydfil. I later returned to college for further education and took a job as a medical secretary. I have had many jobs and, through the years, have always found myself being unsettled in anything I have done. Maybe it is because of my background, who knows? I only wish there was two of me, as then I could make a comparison of the life I might have lived differently if I wasn't connected to the disaster.

After the disaster, as time went by, many groups of people would get together and organise carnivals. These events took our minds off bad times and gave the people of Aberfan something to look forward to. Jazz bands were formed and the carnival events were always a success. Every street had a theme and floats, lorries – anything that moved – were decorated. As a teenager I was picked to be Carnival Queen for Aberfan and Meg Richardson from the *Crossroads* television programme crowned me. Bonfire nights too were an eerie experience. All the village people would march through the streets with torches held high. We would march and then go back to the Grove Field where the fireworks were lit. Aberfan was always like one family.

Not many people talk of the disaster any more. Some people put it right out of their minds, only to discover that years later they have suffered the post-traumatic stress that still has many people in Aberfan on medication.

From the depths of despair that could have destroyed a lesser community, the indomitable spirit of Wales came forth in the lives of 11 men who wanted to help and to rebuild. They began the Aberfan Male Voice Choir, later to be called the Ynysowen Male Voice Choir, and today it helps to enrich the lives of its own community and hundreds of others with its charity work. The choir was formed from the tip removal meetings that members of the community attended. Most of its members were bereaved. Its first annual concert took place in 1967. Trevor Hughes, the oldest choir member, is now 81 years old and has been with the choir since the beginning.

Wales has survived thousands of years of upheaval, turmoil and oppression. It has suffered more than its share of natural and man-made disasters. Yet, its spirit lives: it can be seen in the faces of Ynysowen Male Voice Choir and heard in their voices when they sing.

*

A new school was built in the grounds of Ynysowen, called Ynysowen Primary. It was opened by Harold Wilson. I presented Mr Wilson with a carnation for his jacket. My children attended the school but my mother always found it too difficult to go to any concerts which were held at the school. Hearing and seeing the children perform was too emotional for her to bear. However many long years have passed, there are still scars; the wounds remain unhealed.

Pantglas School was demolished in 1967 and there is now a memorial garden in its place. The shrub borders are laid out to form a plan of every classroom that once stood there. Near the garden, on the site of the houses demolished by the slide, stands a new community centre.

A new group of bereaved parents was formed to run the Aberfan and Merthyr Vale Memorial Garden Committee. Visitors from all over the world still visit the cemetery. I have received many letters from people all around the world who have also experienced trauma; my story has given them hope.

Our cemetery is on a hillside. You can see the line of children's graves even before you enter. They stand out, almost shine. Never forgotten.

As a plaque at the entrance to the memorial garden states: 'To those we love and miss so very much.'

Throughout the years I have always visited the cemetery. I still do so and will never forget. Having children of my own has made the pain even more intense. When my daughter, Cassie, was growing up, I used to look at her sometimes and try to imagine what my parents felt then... to lose two children, in an instant. No words could express my thoughts. Being so young at the time, I never once thought how my parents felt – I can only remember how I felt.

Cassie is now in her late 20s and has two children of her own, Oliver and Jaxson, with her partner Jason. She said recently: 'I can remember on a few occasions asking my nan, Iris, who the little boy and girl were in the big paintings that were hanging in the front room. She would always take a deep swallow and reply to me that they were Carl and Marylyn, but she said nothing else. Although I knew what happened, I never fully understood the consequences and the impact it had on so many people, especially my nan, grampa, mother and aunties. I would never ask my nan what happened as I knew she would get upset. My grampa would talk little bits about it to me and always in a croaky, emotional voice. You knew when to change the conversation.' And of Mam, she said: 'My nan is an unbelievably strong woman and I'm so proud of her. For 49 years she's held everybody together and never given up, and I know my grampa would be very proud of how well she is coping.'

Cassie and her older brothers, my sons James and Ben, are part of the generation which only exists because some of us survived. My three children all grew up in Aberfan, in a family which revolved around my father. They adored him and my mother. The disaster was always in the background for them. James said recently he felt 'fortunate to be here' as part of the first generation since the disaster. He and his wife, Tammy, now have a son, Aaron. James recently confided some of his thoughts to me on paper. In them he talked about how unsettled he realised I had been. He reckoned he had moved 17 times in his life, mainly down to me. 'I don't think my mother will ever settle down properly but she has grown up to become a strong person through pure faith and courage. God has taught us to forgive one another and hopefully parents and survivors can forgive those guilty of the cause of the disaster.'

Hearing from my children as I ended my journey brought tears to my eyes. As I was writing the last few pages of this book, I received this from my second child, Ben. 'I can never do justice in words as to how strong my grandparents have been and how grateful I am that they have passed this strength on to my mother. My mother has always been my hero. The thought of her being defenceless and trapped and afraid is heartbreaking for me, but she carried on with all the emotional scars of that horrific day tearing her apart. She has become an inspirational, happy, strong, independent woman. I have always felt that my mother carried a feeling of guilt with her that she survived and Carl and Marylyn had not. I am proud of you, Mam. You, like Nan and Grandpa are a true inspiration to us all.'

*

Ben has two children with his former partner, Kelly. They are Megan and Lily. Megan is my eldest granddaughter. She is bright, inquisitive and loves spending time down at the local library. She is ten and in the last year at Ynysowen Primary School.

One day I was driving along Aberfan Road with Megan in the front, and Lily, Oliver and Jaxon in the back. They were playing up as usual, and all of a sudden, out of the blue, Megan shouted at top of her voice: 'Shut up, shut up! Stop being naughty and listen to Gaynor! If it wasn't for Gaynor none of us would be here, we wouldn't be alive.'

Her words struck me right in the heart. I had never thought of it like that. I didn't know what to say. I just muttered: 'Megan, I will speak to you about Aberfan another time.'

A few weeks later I was babysitting Megan and Lily, and I wanted to tell Megan some basic information, but also to help her understand what she has been told. Children are often frightened about what is whispered or not mentioned aloud, and I wanted her to understand the truth.

I asked her when she heard about the disaster?

'It was my father. He told me that if it weren't for Gaynor then we would not have been here; then he explained why because I asked him why. He just explained some of the things; he said Gaynor was in that...'

Megan was hesitant to say the word 'disaster'. I asked her what she thought and she said, 'It was horrible. It was terrible.'

I paused, then said: 'You are around the same age as little children who went to heaven that day.'

She nodded. 'When I go to school, sometimes I think something is going to hit us down now.'

'No, Megan,' I said. 'You are safe in your school. It's a brand-new school away from any tips. They will never come back.'

I made sure she understood that. It would never happen here again.

I told Megan about the history of her village, about the Merthyr Vale colliery and what the miners' jobs were, and about the coal that was carried up to be tipped on the mountainside.

Megan had shared some of my writings with a friend who had showed it to his mum. The friend had read my description of the disaster, and he had said: 'I am that age next year and what if I actually die next year? I haven't really lived for long.'

Megan said: 'I think all the kids must have said they wanted to be 16. They wanted a puppy, a mansion, but all their dreams came to a nightmare.'

I said: 'The tips have gone. You and Jack and Lily, enjoy your time playing and school, and go to college to get a good education and job.'

Megan said she had heard about all the mud, about how it had gone through people's doors, had filled the street and the school. 'Did everyone have to go in there to look for their children?'

We looked through some photographs of me as a child.

'Did any of your best friends die?' Megan asked.

'Yes,' I said. 'They are angels now.'

*

The question of whether the Government owed us an apology for the failures of its predecessor and of the NCB came up a number of times on my journey. I liked what James Bullock had to say. He was the soldier who had worked in the morgue after the disaster. 'An apology would have to be from the heart and not just to wipe the slate clean and yet, 50 years on, at whose door do you lay the blame? Those responsible are long dead, but, no, it's never too late to apologise, as long as it doesn't come over as an insult to the people of Aberfan; they deserve much better.'

Thirty years on from the disaster, I wrote in my journal: 'Whoever you are – child or adult – perhaps you too have found yourself in traumatic circumstances during your life. There is light at the end of the tunnel, no matter how long it takes to reach the end. Just have faith, like me, and it will see you through. Above all this suffering, the people of this village have remained strong and proud to be descendants of their Welsh heritage, like our fathers and grandfathers.'

*

I still live in Aberfan. I work for Barnardo's, I just love my job. My dream was always to move away and I tried living in Cardiff for a while but eventually came home. Like my parents had once discovered when they had considered moving away, you cannot escape your past. And Aberfan, the village, the disaster, are instilled in you. There is no running away, it just follows you, and like a magnet the place pulls you back every time.

I am nearly 60 years old now and still single. I have had many relationships but somehow I feel destined to be on my own. When I married at 19 I don't think I was emotionally stable enough. But it gave me three beautiful children. James, Ben and Cassie are now all in their 30s. They have given me five precious grandchildren.

Dad has passed on but Mam is 80 and living in the bungalow that he built in memory of my brother and sister.

My father is not the only person Aberfan has lost as I made this journey. Evelyn Dinnage, the lady who had comforted me on my way to the cemetery when I was a child, died on 17 August 2015. She was 94. I had taken too long to go and see her but I am so grateful that we got to talk in the end. Her kindness and memories had to be recorded; I am so glad I was able to do it.

John Reddy, who had told me all about the education authority's response to the disaster, died on 26 February 2016, aged 82. I will never forget him telling me about the six small white coffins he saw laid out in church.

There is both joy and pain in memory. Every day we have to mark that day. Every day we have to remember the people we lost, and the pleasure their lives brought us.

And because we still live in the community in which the disaster happened, we remain surrounded by memories. Belinda was recently clearing out a chest of drawers in her bedroom when she came across Carl's trouser belt – a simple item packed with emotion. While looking through my parents' cupboards, I came across Marylyn's school books in an old brown leather zipped case. The last entry I could find was an exercise book dated 4 July 1966. On the front she had written her name and class, 3W. The teacher had written a note asking her to improve her spelling.

*

I am aware that I have lived a very roller-coaster life, an unsettled one of doomed personal relationships and an unstable working career. As my children said, I have moved home a lot and suffered periods of anxiety. During the week of the 30th anniversary I had to spend a week in hospital as I suffered terrible panic attacks. I have come to understand that grief is a natural emotion, part of the process which will eventually lead to healing; but trauma and feelings of anger against injustice can block that process. Stop us from healing.

To this day, we still learn new details about the disaster. As I was completing this book Mam was watching a documentary on the Dunblane massacre and she suddenly revealed to me that her and Dad had not had the strength to identify Carl and Marylyn. This had been carried out valiantly by my Auntie Pam and her husband Eddie. They had returned to tell Mam and Dad at 2.30 on the Saturday morning.

For me, there have been many unanswered questions over the last 50 years. I now finally feel able to close the door and put my ghosts to rest.

My situation was unique, and I can't begin to imagine what others are going through in their daily lives. You *can* survive tough situations, is all I can say. I did. History has taught us that even when it looks like there is no hope, that precious emotion still lives in people's hearts.

I am so lucky and blessed for all the wonderful things that I do have and I have struggled all my life to get where I am today. Life is too short to think about all the things you don't have. I have my family and friends, and what more can I ever want? Life is wonderful. Believe me, I know, because I almost lost mine before it had really even begun.

People from all over the world still come to pay their respects at the graves of Aberfan. I met a group of schoolchildren from Cheltenham there recently and afterwards they wrote down their thoughts. 'The statistics learnt will stay with me about the event as I have seen the impact in person,' one wrote. 'Gaynor has made me realise how we need to act more kindly to others, as disasters like this do happen very randomly and can have an impact on anyone.' Another added: 'It made me think of the severe long-term impacts that people deal with all around the world with different natural disasters.'

The village may always be synonymous with a terrible tragedy but it is more than the name of a disaster. For a long time in the village there were no birthdays, engagements or weddings, but now most of my friends have married and all have had children. These children are now teenagers and once more the streets of Aberfan are filled with voices and laughter.

The tears have slowly dried up in Aberfan, but we've only got better little by little. One day at a time.

Appendix

I wrote the following poems when I was aged 12 to 14

Aberfan Friday 21st

Darkness came and the mist was down,
Children sit and learn and frown,
What's that noise? So near, so loud,
Rumble, rumble, frightened faces,
One big crash, all dirt around,
Many lay dead on the ground.

Screaming babies just so young
Some lay bleeding, some lay still,
Cries that filled the air that day,
Someone helps me, I'm so afraid.

Where are my friends? My shoes have gone,
New ones too, my mum will shout.
Should I worry? I'm alive,
I'm in pain, I will survive.

Tears of frustration, anger too,
Questions asked, the hell we went through,
Suffer little children to come unto me,
God has taken them to eternity.

Like the Pied Piper the children were gone,
Leaving the village empty with sorrow,
In our hearts their souls are locked
To reunite in Heaven, when our time comes,
One turn of the key, we shall be one.

Time does heal, faith restored,
Always have hope and you will find
God will reward each of us in kind.

Saved by my Grandpa

Grandpa, I remember when your eyes filled with despair,
Looking at me through broken glass and debris,
Tears I cry, my hands reach out to you,
You look so helpless, you can't get through,
Too many children in your way,
Hurry Grandpa, I am so afraid.

My hand is stuck, my leg is covered,
It takes too long to set me free,
He carries me through all the muck
To my dad who waits anxiously.
Sirens roar, people wander
Through darkness misty morn.
Children buried, some survive
Where's my school, 'It has gone'.

The Village Suffered – Smiles Regained

The village suffered, days, nights and years,
The tips have gone now, we must go on,
Look to the future, no more dark morn.

The sun shines down on us today,
Putting back the happiness where there was dismay,
Green pastures grow, birds sing their song,
Children's laughter is once heard in streets all day long.

God's little angels look down from the skies,
Keeping watch on their families, making them smile,
Don't worry they say, we're safe and sound,
Wrapped up in the wings of the Lord above.

*A tribute poem by Ross Manning, grandson of Mam and Dad;
Belinda's son*

Separated by 50 years

He remembers an age ago, a time buried in the past,
He remembers when everyone went to Church, and nothing
  happened too fast,
He remembers when everyone had a job or were called to fight
  on the sea,
I wish I could remember those times, sometimes I wish I were
  he.

I can't see very far ahead and I don't remember the past,
All I can see is the beginning of the End – maybe I'm one of
  the last,
We're the same young man, he and I, separated by 50 years.
I bet the Valleys he remembers are nothing like the ones here.

He remembers owning one of the few cars in the village – for
  miles around,
Now the cars are choking us up and I have to leave town,
I don't want to leave this place, all my friends and memories,
It's just that I want a better place, somewhere left to breathe.

She remembers bearing six children, and losing two in the
  mud,
She remembers mothering us all, shedding sweat, tears and
  blood,
She remembers marrying him when he was just older than
  me,
I hope I find a woman like her, waiting out there for me.

I know it's too late for this place; they've bled the Valleys dry,
But people like them – they'll keep on going – they won't lie
  down and die!
They're a testament to the Past, the place where I am from,
I hope there will always be people like them, because they'll
  soon be gone.

Poem I wrote when I was 16 after my break-up with David

Summer Breeze

Summer breeze, birds that sing their song,
Blue skies, with pastures green,
Hazy days, nights so long,
Dreaming of thoughts of you, what went wrong?

Spring will come, frosty morns will wake
Telling us seasons change; it's time to make
Promises; no lies, trust and faith,
Wouldn't it be nice to make the world a better place?

Times we shared, beaches we walked,
Holding hands, kissed and talked;
Stolen moments, I hate to leave,
Don't cry, time will tell, feel that summer breeze,
Love will always be there,
Just for us two; we will always share.

A story the Reverend Irving Penberthy told the child survivors of the disaster who attended his Sunday school

Billy Blackbird

Billy Blackbird and Tommy Thrush were the best of friends. They always sang together in the morning and they sang at many other times when they were happy. Tommy would often get carried away by his singing and would close his eyes and lift his head to the skies – but this was very dangerous. When he was young Tommy's mother had often taught him to look left, look right, then left again, before singing to make sure that the coast was clear, but he often forgot.

One day they went down to the village to find some crumbs: Tommy went into one garden and Billy went into the next. The weather was warm and the garden was so full of flowers that

Tommy was overcome with pleasure and, throwing back his head, he began to sing.

But there was a crafty cat around, and cats can do very nasty things. This cat pounced upon Tommy and grabbed him by his claw. Billy heard his cries and came to his rescue. He pecked at the cat's head and ears, shrieking, 'Let go!', until, at last, the cat let Tommy go and skulked away leaving him badly hurt upon the ground.

Billy chirped his distress call and the other birds gathered but there was little that they could do. They sent for the Red Cross Ravens who took Tommy on a stretcher to Doctor Owl who did his best to tend his wounds and treat him for shock.

When Billy went to visit him, Doctor Owl shook his head sadly: 'There is no more I can do. He must go to the Land of Sunshine where the doves will be able to heal him.'

'Where is the Land of Sunshine?' Billy asked.

'It is far beyond the distant mountains. I must send for the Eagle Express.'

The Eagle Express soon arrived with a basket on his back and, when Tommy had been carefully placed in the basket, Billy waved goodbye and the eagle soared into the sky.

When he had gone Billy turned to Doctor Owl and said: 'Please tell me about the Land of Sunshine.'

'It is a very wonderful place,' said Doctor Owl. 'The River of Life flows through the land with trees on its banks and the leaves are for the healing of the nations.'

'Have you been there, Doctor Owl?'

'Not yet, but I will go there one day and so will you,' said Doctor Owl.

'It must be a very beautiful place,' Billy sighed.

'It is. Some call it the Never-Never Land. The sun never sets, the birds never fight and they never grow old.'

'No cats?' said Billy.

'I really don't know the answer to that,' said Doctor Owl. 'If there are cats they will be cats with no claws who only eat cakes.'

Billy looked into the distance and said: 'When will Tommy come back?'

Doctor Owl smiled gently. 'That's another thing about Never-Never Land,' he said. 'They never come back. That is where they belong.' He held his wing out towards Billy comfortingly. 'You can't stay here in the forest for ever, you know.'

Billy asked: 'What shall I do without my dear friend Tommy?'

'Just carry on singing,' said Doctor Owl. 'Tommy will be singing as well by the River of Life... but you must wait your turn.'

So Billy carried on singing in the forest and, somehow, Tommy did not seem so far away.

Once or twice Billy thought he heard Tommy's song on the evening air and he sang more loudly hoping that his friend might hear him too...

... and Billy did not feel so sad any more.

Acknowledgements

THIS BOOK OFFERS a gift of powerful insight, honesty and wisdom into my life and the disaster which impacted on me as a child; and the events that shaped me into the person I am today.

We cannot change the past but we can change the way we feel about it. This book documents the facts and my wish to preserve them.

I dedicate this book to my parents, Cliff and Iris Minett, who were a source of strength to my family, no matter what sort of painful difficulties they were experiencing. Their courage, determination and resilience remain an inspiration to me and my family. Bless you both dearly.

My family is my universe. They have given me so much support, not only through the process of writing this book, but throughout my life. My three sisters, Belinda, Michele and Sian, through 50 years they have never left my side. We have laughed and cried together. I will always treasure their support and love. They gave me encouragement when it seemed too difficult to complete the book. I would have probably given up without their support.

I would like to pay special tribute to my three beautiful children, James, Ben and Cassie. I am so proud of you all. I have watched you grow, bringing laughter, tears and sleepless nights, and I am now blessed and honoured to be called Grandma by five beautiful grandchildren: Megan and Lily Madgwick, Oliver and Jaxon Rudge, and the newest addition to the family, Aaron Madgwick. You are my little treasures, I love you.

This book has been an emotional journey, not just for me, but for the many people who have spoken to me and shared their stories and research material. Their honesty and candour have allowed me to gain insight into the Aberfan disaster which I could have not done alone. Many thanks to you all from the bottom of my heart; without your contributions this book may not have been possible. A huge thank you: you have contributed to ensuring valuable lessons and insight is available to all.

I must give thanks to Professor Iain Mclean, in particular, for his support for the book; for immediately connecting with my story, giving up his valuable time and travelling many hours to meet me. Dr Martin Johnes, his colleague, also gave me direction and valuable research information which I found so useful.

Many thanks also to Greg Lewis who worked to turn my interviews into a book – without your support the book would not have been written. I am very grateful. And also to our agent Jeffrey Simmons for immediately connecting with my story.

Mel Doel accompanied me to a number of the interviews. She has been hugely supportive throughout my journey, generous with her time and there when I needed advice; she has been a true friend of the family for many years. She was able to contact legendary broadcaster Vincent Kane, her colleague for many years at the BBC. It was an honour to meet Vincent during the writing of this book. He is a true gentleman. I would like to thank him immensely for his time and support, and for writing a truly powerful introduction – a penetrating insight from a great observer of Aberfan's politics.

Meeting Lord Snowdon after 50 years was emotional. How thankful and appreciative I am for an amazing moment in my life. Thank you for taking the time to write the foreword and sharing your empathy towards me and the people of Aberfan. I am also grateful to Lord Snowdon's PA, Lynne, for being so responsive to my communications and making my visit to his home possible.

A special thank you to Ron Davies who, for many years, had

been at the forefront of Dad's thoughts; he has always been highly spoken of in our family home. He, along with others, ensured financial justice for Aberfan. Thank you for sharing your story, an emotional interview. Dad can now rest in peace, no more 'fighting'. Thank you.

Finally, this book is a tribute to Dad. The hero, the fighter. The core of our inspiration. Days from entering Heaven you asked me to finish the book. I will never forget those words. As you now rest in peace, I kept my promise to you, Dad. Here is the book, your book. I did it just for you. One day at a time. Thank you for believing in me, Dad.

Gaynor Madgwick
Aberfan
February 2016